Novel Trends in the Traveling Salesman Problem

*Edited by Donald Davendra
and Magdalena Bialic-Davendra*

Published in London, United Kingdom

IntechOpen

Supporting open minds since 2005

Novel Trends in the Traveling Salesman Problem
http://dx.doi.org/10.5772/intechopen.78197
Edited by Donald Davendra and Magdalena Bialic-Davendra

Contributors
Weiqi Li, Donald Davendra, Magdalena Bialic-Davendra, Magdalena Metlicka, Fernando Francisco Sandoya, Carmen Letamendi, Fanny Sanabria, Wayne Frasch, Michael Kuby, Fusheng Xiong

Notice
Statements and opinions expressed in the chapters are these of the individual contributors and not necessarily those of the editors or publisher. No responsibility is accepted for the accuracy of information contained in the published chapters. The publisher assumes no responsibility for any damage or injury to persons or property arising out of the use of any materials, instructions, methods or ideas contained in the book.

First published in London, United Kingdom, 2020 by IntechOpen
IntechOpen is the global imprint of INTECHOPEN LIMITED, registered in England and Wales, registration number: 11086078, 5 Princes Gate Court, London, SW7 2QJ, United Kingdom
Printed in Croatia

British Library Cataloguing-in-Publication Data
A catalogue record for this book is available from the British Library

Additional hard and PDF copies can be obtained from orders@intechopen.com

Novel Trends in the Traveling Salesman Problem
Edited by Donald Davendra and Magdalena Bialic-Davendra
p. cm.
Print ISBN 978-1-83962-453-7
Online ISBN 978-1-83962-454-4
eBook (PDF) ISBN 978-1-83962-455-1

We are IntechOpen,
the world's leading publisher of
Open Access books
Built by scientists, for scientists

5,100+
Open access books available

126,000+
International authors and editors

145M+
Downloads

156
Countries delivered to

Our authors are among the
Top 1%
most cited scientists

12.2%
Contributors from top 500 universities

CLARIVATE ANALYTICS
BOOK
CITATION
INDEX
INDEXED

WEB OF SCIENCE™

Selection of our books indexed in the Book Citation Index
in Web of Science™ Core Collection (BKCI)

Interested in publishing with us?
Contact book.department@intechopen.com

Meet the editors

Dr. Donald Davendra is Professor and Chair of Computer Science at Central Washington University, USA. He has a PhD in Technical Cybernetics from Tomas Bata University in Zlín, Czech Republic. His research background is in optimization algorithms, intelligent manufacturing systems, high-performance computing, and chaos systems. He has edited three books and numerous journal papers, book chapters, and conference publications in the field of computational intelligence.

Dr. Magdalena Bialic-Davendra has a PhD in Finance from Tomas Bata University in Zlín, Czech Republic. She is author and co-author of more than thirty-five scientific articles. Her research interests include wildfire modeling, business clusters, operations research, and management science.

Contents

Preface

Dedicated to Kinga

The Traveling Salesman Problem (TSP) is widely considered one of the most intensively studied problems in computational mathematics and operations research. Since its inception in the 1800s, it has become the poster child for computational complexity research and Graph Theory. A number of problems have been transformed to a TSP problem, and its application base extends into scheduling, manufacturing, routing, and logistics, among others. With the advent of high-performance computing and advanced meta-heuristics such as Graphical Processing Unit (GPU) programming and Swarm-based algorithms, the TSP problem is positioned firmly as the go-to problem in the development of the next generation of intelligent heuristics.

This book is targeted to students and researchers. It encompasses the latest trends in TSP applications, including both theory and practical aspects, with emphasis on cutting-edge algorithms that incorporate unique paradigms such as high-performance computing using GPUs, software accelerators, and meta-heuristics.

Donald Davendra
Department of Computer Science,
Central Washington University,
Ellensburg, USA

Magdalena Bialic-Davendra
Department of Economics and Department of
Finance & Supply Chain Management,
Central Washington University,
Ellensburg, USA

Chapter 1

Introductory Chapter: Traveling Salesman Problem - An Overview

Donald Davendra and Magdalena Bialic-Davendra

1. Introduction

The traveling salesman problem (TSP) is considered one of the seminal problems in computational mathematics. Considered as part of the Clay Mathematics Institute Millennium Problem with its assertion of $\mathcal{P} = \mathcal{NP}$ [1], the TSP problem has been well researched during the past five decades.

The TSP problem can be described as the following: consider a number of cities which must be visited by a traveling salesman, only once, arriving once and departing once and starting and ending at the same city. Given the pairwise distances between cities, what is the best order in which to visit them, so as to minimize the overall distance traveled?

Mathematically, the equation for the TSP can be given as in Eq. (1):

$$x_{ij} = \begin{cases} 1 & \text{the path goes from city } i \text{ to city } j \\ 0 & \text{otherwise} \end{cases} \tag{1}$$

where $x_{ij} = 1$ if city i is connected with city j, and $x_{ij} = 0$ otherwise. For $i = 0, \dots, n$, let u_i be an artificial variable and finally take c_{ij} to be the distance from city i to city j. The objective function can be then formulated as Eq. (2):

$$
\begin{aligned}
&\min \sum_{i=0}^{n} \sum_{j \neq i, j=0}^{n} c_{ij} \, x_{ij} \\
&0 \leq x_{ij} \leq 1 && i, j = 0, \dots, n \\
&u_i \in \mathbb{Z} && i = 0, \dots, n \\
&\sum_{i=0, i \neq j}^{n} x_{ij} = 1 && j = 0, \dots, n \\
&\sum_{j=0, j \neq i}^{n} x_{ij} = 1 && i = 0, \dots, n \\
&u_i - u_j + n x_{ij} \leq n - 1 && 1 \leq i \neq j \leq n
\end{aligned}
\tag{2}
$$

2. Complexity

The complexity of the TSP is still unknown. Using a brute force approach to test each and every tour, for a tour of n cities, it will be (n−1)! possibilities and it will have a time complexity of $O(n!)$. However, using the dynamic programming approach, the complexity can be derived of a tour of n cities, which can be divided into n−2 subsets each of size n−1, with the constraint that all such subsets don't have the nth city in them. Therefore, there are a maximum of $O(n2^n)$ such subproblems, which can be solved in *linear* time. The time complexity is therefore $O(n^2 2^n)$. Both space and time complexity of the TSP problem can be considered as *exponential*.

3. History

The genesis of the TSP problem is difficult to pinpoint. Some literature point to widespread usage since the 1950's [2], after the *48 state problem* posed by Hassler Whitney in the 1930's induced a lot of interest. The subsequent second world war really ingrained the use of operations research into this domain. An excellent detailed history is given in [3], where TSP is considered as a part of the history of Combinatorial Optimization.

The TSP problem over time has evolved into many different branches, each with different constraints:

Symmetric TSP (STSP) - the basic TSP problem, where the distance between city i and city j is the same as from city j and city i.

Asymmetric TSP (ATSP) - modified TSP, where the distance between city i and city j is *not* the same as from city j and city i.

Hamiltonian Cycle Problem (HCP) - is a problem where finding if a path in an *undirected* or *directed* graph G that visits each vertex V exactly once exists.

Sequential Ordering Problem (SOP) - Given a set of n cities and distances for each pair of cities, find a *Hamiltonian path* from city *1* to city n of minimal length, which takes given precedence constraints (such as requiring some nodes to be visited prior) into account.

Capacitated Vehicle Routing Problem (CVRP) - Given *n-1* nodes, 1 depot (with resources) and distances between the nodes, the objective is to satisfy demand at each node using vehicles with identical capacity with the shortest tour.

Case Enough TSP (CETSP) - a problem developed for radio frequency identification (RFID), where close proximity is enough to a node. The objective is to trace the shortest path interlinking the different nodes.

TSP with Neighborhoods (TSPN) - where a collection of $\mathscr{L}$ regions in the plane, called *neighborhoods* is given and the objective is to seek the shortest tour to visit all these neighborhoods.

4. Current literature

Linear programming and deterministic methods have been seen as the early solvers, however, intractability of this problem has led to a general decline in these mathematical formulations. Within the past few decades with the rise of computational power, a new branch of algorithms called *meta-heuristics* generally based on evolutionary dynamics have become more synonymous with solving combinatorial optimization problems. Based around naturally occurring phenomena, these algorithms treat each problem as a black box with the aim of finding feasibly good solutions within acceptable space and time constraints. A vast repository of literature exists for the TSP problem, and the TSP Library is an excellent starting off resource point [4].

4.1 Deterministic approaches

Some of the latest literature on the TSP problem is divided into three components. The first is the exact and approximation algorithms, which try and produced efficient and reasonably good quality solutions. Some of the latest approaches are given below.

1. 2-Opt Algorithm [5]

2. Branch and Cut Algorithm [6]

3. Approximate and Exact Algorithms [7]

4. Branch and Bound [8]

4.2 Evolutionary approaches

The second aspect is evolutionary algorithms. A vast number of these algorithms are now in existence and have been applied to the TSP problem from the seminal work on the Ant Colony Optimization by Dorigo and Gambardella [9] to the following current research.

1. Artificial Bee Colony [10]

2. Differential Evolution [11]

3. Genetic Algorithm [12]

4. Tree Seed Algorithm [13]

5. Spider Monkey [14]

6. Ant Colony Optimization [15]

7. Harmony Search Algorithm [16]

8. Pigeon Inspired Optimization [17]

4.3 High performance computing

The third aspect is application based, specifically high-performance computing. With the wider dissemination of parallel computing, especially multi-core and graphic processor unit based approaches, many algorithms have been parrallized. Some of the latest approaches from literature is given as:

1. Multi-Core approach [18]

2. OpenMP [19]

3. CUDA [20]

5. Future direction

Even though a number of problems, especially in the combinatorial and scheduling domain have increased over the past decade, the TSP problem have remained a vital area of research. This is primarily for it being generally equated to the *intractably* quandary of $\mathscr{P} = \mathscr{N}\mathscr{P}$, with its far reaching consequences in other fields such as encryption etc. It is the belief that a combination of smart heuristics employed on super-computers with parallel programming paradigms will be the future direction of tacking large-scale TSP problems.

Author details

Donald Davendra*† and Magdalena Bialic-Davendra†
Central Washington University, Ellensburg, USA

*Address all correspondence to: donald.davendra@cwu.edu

† These authors contributed equally.

IntechOpen

References

[1] Clay Mathematics Institute https://www.claymath.org/millennium-problems/p-vs-np-problem [Accessed: 10 October 2020]

[2] Applegate DL, Bixby RE, Chvatal V, Cook WJ. The Traveling Salesman Problem: A Computational Study. Princeton. Oxford: Princeton University Press; 2006

[3] Alexander S. On the History of Combinatorial Optimization (Till 1960), Editor(s): K. Aardal, G.L., Nemhauser, R., Weismantel, Handbooks in Operations Research and Management Science, Elsevier, Vol 12, Pages 1-68, 2005

[4] TSP Library. http://comopt.ifi.uni-heidelberg.de/software/TSPLIB95/ [Accessed: 10 October 2020]

[5] Hougardy S, Zaiser F, Zhong X. The approximation ratio of the 2-Opt Heuristic for the metric Traveling Salesman Problem. Operations Research Letters. 2020;**48**(4):401-404

[6] Yuan Y, Cattaruzza D, Ogier M, Semet F. A branch-and-cut algorithm for the generalized traveling salesman problem with time windows. European Journal of Operational Research. 2020;**286**(3):849-866, ISSN 0377-2217. DOI: 10.1016/j.ejor.2020.04.024

[7] Wang S, Liu M, Chu F. Approximate and exact algorithms for an energy minimization traveling salesman problem. Journal of Cleaner Production. 2020;**249**:119433, ISSN 0959-6526. DOI: 10.1016/j.jclepro.2019.119433

[8] Salman R, Ekstedt F, Damaschke P. Branch-and-bound for the Precedence Constrained Generalized Traveling Salesman Problem. Operations Research Letters. 2020;**48**(2):163-166, ISSN 0167-6377. DOI: 10.1016/j.orl.2020.01.009

[9] Dorigo M, Gambardella L. Ant colony system: a cooperative learning approach to the traveling salesman problem. IEEE Transactions on Evolutionary Computation. April 1997; **1**(1):53-66. DOI: 10.1109/4235.585892

[10] Pandiri V, Singh A. An artificial bee colony algorithm with variable degree of perturbation for the generalized covering traveling salesman problem. Applied Soft Computing. 2019;**78**: 481-495, ISSN 1568-4946. DOI: 10.1016/j.asoc.2019.03.001

[11] Ali I, Essam D, Kasmarik K. A novel design of differential evolution for solving discrete traveling salesman problems. Swarm and Evolutionary Computation. 2020;**52**:100607, ISSN 2210-6502. DOI: 10.1016/j.swevo.2019.100607

[12] Dong X, Cai Y. A novel genetic algorithm for large scale colored balanced traveling salesman problem. Future Generation Computer Systems. 2019;**95**:727-742, ISSN 0167-739X. DOI: 10.1016/j.future.2018.12.065

[13] Cinar A, Korkmaz S, Kiran M. A discrete tree-seed algorithm for solving symmetric traveling salesman problem. Engineering Science and Technology, an International Journal. 2020;**23**(4): 879-890, ISSN 2215-0986. DOI: 10.1016/j.jestch.2019.11.005

[14] Akhand MAH, Ayon I, Shahriyar SA, Siddique N, Adeli H. Discrete Spider Monkey Optimization for Travelling Salesman Problem. Applied Soft Computing. 2020;**86**: 105887, ISSN 1568-4946. DOI: 10.1016/j.asoc.2019.105887

[15] Tuani A, Keedwell E, Collett M. Heterogenous Adaptive Ant Colony Optimization with 3-opt local search for the Travelling Salesman Problem. Applied Soft Computing. 2020;**106720,**

ISSN 1568-4946. DOI: 10.1016/j.
asoc.2020.106720

[16] Boryczka U, Szwarc K. The
Harmony Search algorithm with
additional improvement of harmony
memory for Asymmetric Traveling
Salesman Problem. Expert Systems with
Applications. 2019;**122**:43-53, ISSN
0957-4174. DOI: 10.1016/j.
eswa.2018.12.044

[17] Zhong Y, Wang L, Lin M, Zhang H.
Discrete pigeon-inspired optimization
algorithm with Metropolis acceptance
criterion for large-scale traveling
salesman problem. Swarm and
Evolutionary Computation. 2019;**48**:
134-144, ISSN 2210-6502. DOI: 10.1016/
j.swevo.2019.04.002

[18] Wei X, Ma L, Zhang H, Liu Y.
Multi-core-, multi-thread-based
optimization algorithm for large-scale
traveling salesman problem. Alexandria
Engineering Journal. 2020. DOI:
10.1016/j.aej.2020.06.055

[19] Burkhovetskiy V, Steinberg B.
Parallelizing an exact algorithm for the
traveling salesman problem. Procedia
Computer Science. 2017;**119**:97-102,
ISSN 1877-0509. DOI: 10.1016/j.
procs.2017.11.165

[20] Ermis G, Catay B. Accelerating local
search algorithms for the travelling
salesman problem through the effective
use of GPU. Transportation Research
Procedia. 2017;**22**:409-418, ISSN
2352-1465. DOI: 10.1016/j.
trpro.2017.03.012

Chapter 2

CUDA Accelerated 2-OPT Local Search for the Traveling Salesman Problem

Donald Davendra, Magdalena Metlicka
and Magdalena Bialic-Davendra

Abstract

This research involves the development of a compute unified device architecture (CUDA) accelerated 2-opt local search algorithm for the traveling salesman problem (TSP). As one of the fundamental mathematical approaches to solving the TSP problem, the time complexity has generally reduced its efficiency, especially for large problem instances. Graphic processing unit (GPU) programming, especially CUDA has become more mainstream in high-performance computing (HPC) approaches and has made many intractable problems at least reasonably solvable in acceptable time. This chapter describes two CUDA accelerated 2-opt algorithms developed to solve the asymmetric TSP problem. Three separate hardware configurations were used to test the developed algorithms, and the results validate that the execution time decreased significantly, especially for the large problem instances when deployed on the GPU.

Keywords: traveling salesman problem, CUDA, 2-opt, local search, GPU programming

1. Introduction

This research addresses two very important aspects of computational intelligence, algorithm design, and high-performance computing. One of the fundamental problems in this field is the TSP, which has been used as a poster child for the notorious $\mathscr{P} = \mathscr{N}\mathscr{P}$ assertion in theoretical computer science.

TSP in nominal form is considered *NP-Complete*, when attempted using exact deterministic heuristics. The time complexity when solving it using the *Held-Karp algorithm* is $O(n^2 2^n)$ and the space complexity is $O(n2^n)$. When solving the problem using optimization algorithms and approximation, then problem tends to be *NP-Hard*.

2-opt is considered the simplest local search for the TSP problem. Theoretical knowledge about this heuristic is still very limited [1]; however, simple euclidean distance variants have been shown to have complexity of $O(n^3)$ [2]. Generally, the computed solution has been shown to be within a few percentage points of the global optimal [3].

One of the empirical approaches of improving the execution of the algorithm is applying high performance computing (HPC) paradigm to the problem. This is generally possible if the problem is deducible to a parallel form.

A number of different HPC approaches exist, namely, *threads*, *OpenMP*, *MPI* and *CUDA*. CUDA is by far the most complex and accelerated approach, as it requires programming on the GPU instead of the central processing unit (CPU).

Since its inception, CUDA has been widely used to solve a large number of computational problems [4]. This research looks to harness this approach to implement the 2-opt approach to the TSP problem.

The outline of the chapter follows with the introduction of the mathematical background of the TSP problem followed by the 2-opt algorithm. CUDA is subsequently discussed and the two CUDA developed 2-opt algorithm variants are described. The experimentation design discusses the hardware specifications of the three different architectures and then the obtained results are discussed and analyzed in respect to the execution time.

2. Traveling salesman problem

The TSP is a well-studied problem in literature [5, 6], which in essence tries to find the shortest path that visits a set of customers and returns to the first. A number of studies have been done using both approximation-based approaches [7] and metaheuritics. Metaheuritics are generally based on evolutionary approaches. A brief outline of different approaches can be obtained from:

1. Tabu Search: [8]

2. Simulated Annealing: [9]

3. Genetic Algorithm: [10, 11]

4. Ant Colony Optimization: [12]

5. Particle Swarm Optimization: [13]

6. Cuckoo Search: [14]

7. Firefly Algorithm: [15]

8. Water Cycle Algorithm: [16]

9. Differential Evolution Algorithm: [17]

10. Artificial Bee Colony: [18]

11. Self Organizing Migrating Algorithm: [19]

The TSP function can be expressed as shown in Eq. (1).

$$x_{ij} = \begin{cases} 1 & \text{the path goes from city } i \text{ to city } j \\ 0 & \text{otherwise} \end{cases} \tag{1}$$

where $x_{ij} = 1$ if city i is connected with city j, and $x_{ij} = 0$ otherwise. For $i = 0, \ldots, n$, let u_i be an artificial variable and finally take c_{ij} to be the distance from city i to city j. The objective function can be then formulated as Eq. (2):

$$\min \sum_{i=0}^{n} \sum_{j \neq i, j=0}^{n} c_{ij} \, x_{ij}$$

$$0 \leq x_{ij} \leq 1 \qquad\qquad\qquad i, j = 0, \dots, n$$

$$u_i \in Z \qquad\qquad\qquad i = 0, \dots, n$$

$$\sum_{i=0, i \neq j}^{n} x_{ij} = 1 \qquad\qquad\qquad j = 0, \dots, n \qquad\qquad (2)$$

$$\sum_{j=0, j \neq i}^{n} x_{ij} = 1 \qquad\qquad\qquad i = 0, \dots, n$$

$$u_i - u_j + n x_{ij} \leq n - 1 \qquad 1 \leq i \neq j \leq n$$

3. 2-OPT algorithm

The 2-opt algorithm is one of the most famous heuristics developed originally for solving the TSP problem. It was first proposed by Croes [20]. Along with 3-opt, generalized as k-opt [21], these heuristics are based on exchange of up to k edges in a TSP tour (more information on application of k-opt local search techniques to TSP problems can be obtained from [22]). Together they are called exchange or local improvement heuristics. The exchange is considered to be a single move, from this point of view, such heuristics search the neighborhood of the current solution, that is, perform a local search and provide a locally optimal solution (k-optimal) to the problem [23].

The 2-opt procedure requires a starting feasible solution. It then proceeds by replacing the two non-adjacent edges, (v_i, v_{i+}) and (v_j, v_{j+}) by (v_i, v_j) and (v_{i+}, v_{j+}), and reversing one of the subpaths produced by dropping of edges, in order to maintain the consistent orientation of the tour. For example, the subpath $(v_i, v_{i+}, \dots, v_j, v_{j+})$ is replaced by $(v_i, v_j, \dots, v_{i+}, v_{j+})$. The solution cost change produced in this way can be expressed as $\Delta_{ij} = c(v_i, v_j) + c(v_{i+}, v_{j+}) - c(v_i, v_{i+}) - c(v_j, v_{j+})$. If $\Delta_{ij} < 0$, the solution produced by the move improves upon its predecessor. The procedure iterates until no move where $\Delta_{ij} < 0$ (no improving move) can be found [24].

The 2-opt local search was described by Kim et al. [25] as follows:

Step 1: Let S be the initial solution, $f(S)$ its objective function value. Set $S^* = S, i = 1, j = i + 1 = 2$.

Step 2: Consider exchange result S' such that $f(S') < f(S^*)$. Set $S^* = S'$. if $j < n$ repeat *step 2*. Otherwise set $i = i + 1$ and $j = i + 1$. if $i < n$ repeat *step 2*, otherwise go to *step 3*.

Step 3: if $S \neq S^*$ set $S = S^*, i = 1, j = i + 1$ and go to *step 2*. Otherwise output best solution S and terminate the process.

4. CUDA

General purpose GPU computing (GPGPU) programming was introduced by Apple Cooperation, which created the Kronos Group [26] to further develop and promote this new approach to accelerate scientific computing paradigms.

GPU's offer significantly faster acceleration due to their uniquer hardware archi-
tecture. GPGPU's started to increase in application from 2006. At this point
NVIDIA decided to create its propriety unique architecture called Compute Unified
Device Architecture (CUDA), specific for their *Tesla* generation GPU cards. In
order to support this architect, specific API primitive extensions of C, C++ and
Fortran extensions has been developed [27, 28].

The specific C/C++ language extension for the C language is called the
CUDA-C. This contains a number of accelerated libraries, extensions, and APIs.
These are scalable and freely available without professional license. The main
computational bottleneck is the splitting of the task between GPU and CPU tasks,
where CPU handles better memory management and memory checking and GPU
handles the data acceleration using parallization. It is considered *heterogenous*
programming, where compute intensive data parallel tasks are offloaded on to
the GPU.

CUDA contains three specific paradigms, *thread hierarchy*, *memory hierarchy* and
synchronization. These can be further divided into *coarse-grained* parallelism on the
blocks in *grid* parallization and *fine-grain* parallization in the *threads* in *block*, which
requires low-level synchronization.

4.1 Thread hierarchy

CUDA kernels are special function calls, which is used for data parallization.
Each kernel launches *threads* which are grouped into *blocks* which are then grouped
into *grids*. Communication is done synchronously by *threads* in a *block*, whereas
blocks are independent. Certain programming techniques needs to be undertaken to
ensure data synchronization and validity between *blocks*. *Threads* in different *blocks*
are not able to communicate with each other.

Threads are distinguished by their unique *threadId* in their respective *blockId*,
which allows operating on specific data in the *global* and *shared* memory.

4.2 Memory hierarchy

There are different memory types in the GPU, which CUDA can utilize. Some
memory structures are based on *cache*, some are read-only, etc. The first higher
level memory structure is called the *global memory*, which can be accessed by all
memory *blocks*. Due to its size and access level, it is the slowest memory on the
GPU. The second memory level is the *shared memory*, which is shared by blocks,
which threads within blocks can access. The third memory is the register memory,
which are only accessible by threads, and can be used to local variables. This is the
smallest and fastest memory in the GPU. If there are larger memory structures, and
when registers are not sufficient, *local* memory can be then utilized. Another memory is *constant* memory which cannot be changed by the kernel code. The final
memory is the *texture* memory, which is a read-only cache that provides a speed-up
for locality in data access by threads [29].

4.3 Synchronization

Blocks in *grids* are used in coarse-grained parallelism and *threads* in a specific
block are used in *fine-grained* parallelism. Data sharing in the scope of a kernel is
done by *threads* in the *block*. The number of *threads* are limited by the device
architecture design (max. 1024) and also by *thread* memory resource consumption.
There is a level of scalability as the *blocks* are scheduled independently. Each *block* is
assigned to a streaming multiprocessor (MS) in the GPU [29, 30].

5. CUDA-based 2-opt algorithm

This section presents the parallel CUDA-based version of 2-opt algorithm. This is a modification of the local search for permutative flowshop with makespan criterion problem [31] and its NEH variant [32]. Before coming to the parallel implementation description, however, the more detailed pseudocode of sequential version is provided in Algorithm 5, in order to enable better understanding of the CUDA algorithm design.

Algorithm 1: 2-opt sequential version. The *Swap(T,j,i)* procedure swaps *j*−th and *i*−th cities of tour *T*

Input:
S : initial solution

1
  // number of cities
2 $N \leftarrow Size(S)$
  // objective function value of S
3 $f_S \leftarrow f(S)$
  // temporary solution memory
4 $T \leftarrow S$
5 **while** *ImprovementFound* **do**
6   *ImprovementFound* $\leftarrow$ False
7   **for** *i=1* **to** *N-1* **do**
8   **for** *j=i+1* **to** *N* **do**
9     $T \leftarrow Swap(T,i,j)$
10     $f_T \leftarrow f(T)$
11   **if** $f_T \; ; \; f_S$ **then**
12      $S \leftarrow T$
13      $f_S \leftarrow f_T$
14   *ImprovementFound* $\leftarrow$ True
15   break(2)
16   **end**
17     $T \leftarrow Swap(T,j,i)$
18   **end**
19   **end**
20 **end**
21 return S

As can be seen already from the analysis of description of 2-opt, the task that can be done in parallel is the exploration of neighborhood of the current solution. This is divided between individual CUDA blocks. Possible neighbors of the current solution are split evenly between the launched *blocks*, which then explores these neighbors evenly including the fitness evaluations. If a new better solution is found, it is then stored into the *global memory* allocation of that block. Thereafter, if at least one of the launched *blocks* finds an improving solution during the iteration, the best cost solution amongst all *blocks* is obtained and stored into memory as the current solution for the next iteration. Otherwise, the current solution is returned as the best. It should be noted that the fitness function is not parallelized, as only a single *thread* in each *block* is tasked with this task.

Each *block* explores approximately the same amount of possible neighbors to the current solution (in the worst case, when no improving solution is found), including the cost evaluation. However, if it finds an improving solution, that solution is stored into the *global memory* allocated for each *block*, and the *block* terminates. If at least one of the *blocks* found an improving solution, the minimal cost solution amongst all *blocks* is found and stored into memory as the current solution for the next iteration. Otherwise, the current solution is returned. The cost function evaluation itself was not parallelized, as in each *block* only a single *thread* performs this task.

The outline of the parallel algorithm can be given as follows:

Step 1: Set current solution S = Initial solution.
Step 2: Explore the neighborhood of S by G blocks in parallel. In each block b:
> **Step 1.1:** Determine initial index i for b.
> **Step 1.2:** Explore all neighbors of S created by swapping of i and $j, j \in \{1, \ldots, N\}$. If improving neighbor T found, go to *step 1.4*.
> **Step 1.3:** Determine next index i for b. If $i \geq N$, terminate. Otherwise go to *step 1.2*.
> **Step 1.4:** Store T and its objective function value f_T into global memory and terminate.

Step 3: If no improving solution found, exit procedure and return S as the best solution found. Otherwise determine the best solution amongst those found by blocks in parallel.
Step 4: Store best solution as S. Go to *step 2*.

Where N is the number of cities in the tour and i is the outer loop index (see Algorithm 1 for sequential version of 2-opt).

5.1 Exploration and evaluation of neighboring solutions

The neighbors of solution are generated and evaluated in this kernel. From the sequential version pseudocode (Algorithm 1), it is obvious that the function of generating individual neighbors by swapping every possible pair of jobs pair-wise (i, j) for $i = 1, \ldots, N$ and $j = i + 1, \ldots, N$ can be considered independent and therefore executed in parallel. These solutions can be stored in the *shared memory* after generation. After evaluation, if the new solution has better fitness value compared to the current one, it is stored into the *global memory* allocated for each *block*, to avoid data races between *blocks* (this is illustrated in **Figure 1** depicting memory layout for six cities and four blocks). The improvements counter in the *global memory* is incremented using an atomic operation. This counter is compared against zero after the kernel termination, to determine if the stopping criterion of the algorithm was met. The fitness function itself is evaluated by only a single *thread*; the other *threads* in a *block* process the elements of the solution when transferring data between *shared* and *global memory* locations.

It is logically impractical to allocate the full number of $(N - 1)^2 / 2$ *blocks* on the GPU in most case scenarios. This number can be very large, whereas the number of SMs and the number of resident *blocks* on SM is limited by various factors, such as the number of *threads* in a *block* and a *registers/shared memory* usage.

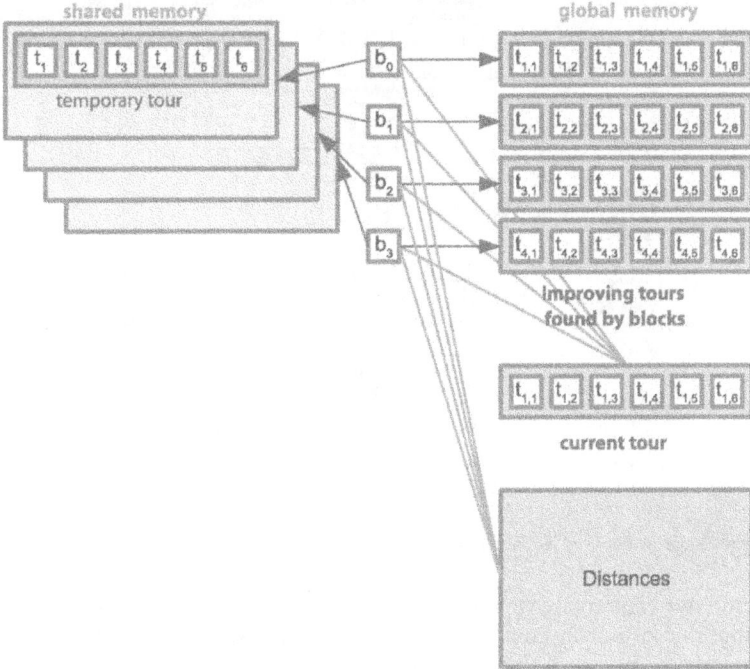

Figure 1.
CUDA-based 2-opt memory layout.

The optimal number of *threads* in a *block* maximizing the number of resident *blocks*, as well as GPU occupancy, can be easily determined based on the calculations performed in the CUDA occupancy calculator tool [33], as a function of the number of cities in a tour (which determines the size of *shared memory* used). This can maximizes the utilization of the GPU, while reducing the total *global memory* size required by the *grid*, as well as the workload done by the search for minimal cost solution in the next kernel. The mapping of the *blocks* to the tasks however can becomes more complicated to implement in code.

Using the assumption that the number of *blocks* will be nearly always smaller than the aforementioned function of the number of actual cities for the problem instances of interest (problems with cities larger than 30), only the outer loop of the sequential 2-opt algorithm was parallelized. The inner loop is performed by each *block* sequentially. This reduces the data transfers between *global* and *shared memory*, and does not eliminate the advantage of the low complexity of the swap operation at the same time. If the solution created by swapping jobs i and j is worse than the current one, it is easy to reverse this change by swapping again j and i, with equal complexity. Therefore, maximally $N - 1$ *blocks* are needed for this function. The mapping of *blocks* to tasks is illustrated in **Figure 2**.

5.2 Parallel reduction to obtain minimal cost

The *parallel reduction* procedure is used to find the index of the solution with the minimal fitness value. This employs *shared memory* to store the data being used, whereas the data is initially copied from the *global* to *shared memory*. In this step, each active thread compares two costs, and stores the smaller of the two costs on the place of the first cost, along with its original index (cost is represented as a structure

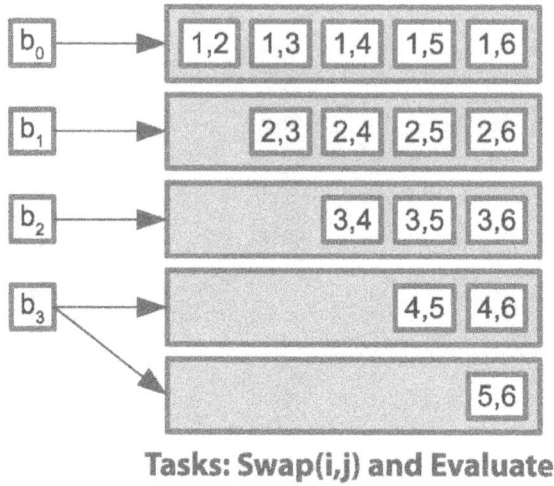

Tasks: Swap(i,j) and Evaluate

Figure 2.
CUDA-based 2-opt, mapping of blocks to tasks.

containing two elements: cost value, and cost index). Using this reduction, the first element of the costs array contains the minimal cost found, along with its respective solution index. This pair is then written into *global memory*.

5.3 Device synchronization and subsequence update

In the final process, a new kernel copies the best indexed solution into the current solution buffer, and the next step of the main loop can be performed. A global CUDA device synchronization is required for relatively large data (for a tour size/number of *threads* in a *block* of size more than approximately 100, as was empirically confirmed) before the start of the synchronization. As each of the

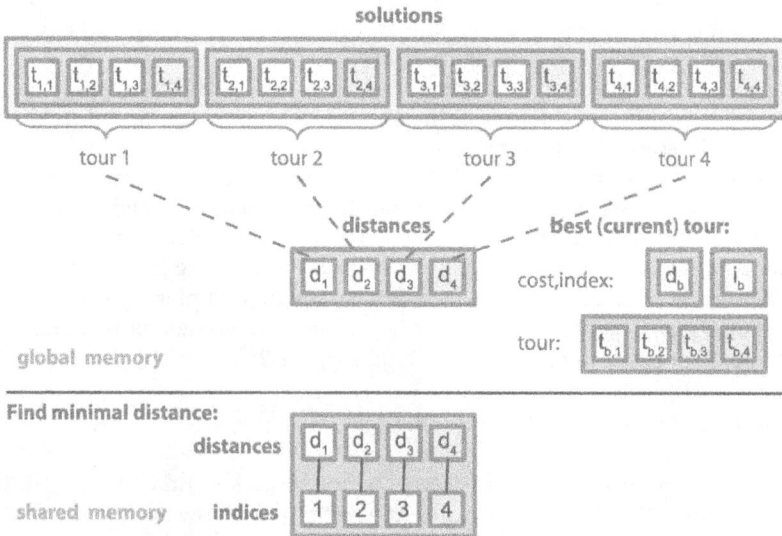

Figure 3.
CUDA-based 2-opt distance and indices layout.

kernels consumes some of the GPU resources, it is necessary to wait, until the pending kernels completely finish the execution, and release their resources, otherwise the GPU freezes and unsuccessful kernel launches start to appear. This is done by calling cudaDeviceSynchronize() function from the host code, after the Update kernel is launched.

Figure 3 outlines the memory layout of the previously described code (without TSP input data, for the current subsequence size 2, city tour 4. The data fields not used in the current step are grayed out). The candidate solutions are stored in one *global memory* 1D array, which conceptually represents 2D array, wherein each row contains one candidate tour. The respective costs are stored in a separate array. The TSP problem input data (distance between cities) are stored in the similar fashion in *global memory* (because of its large size).

This implementation is expected to provide in each step the speedup proportional to the number of solutions generated.

6. 2-OPT variants

Two versions of the 2-opt local search was implemented in this work. The first is the **LS2OPT** variant, which uses the search with the *first ascend strategy*. In this strategy, the next tour is the *first* improving solution found. This can be given in Algorithm 2.

The second variant is the **MLS2OPT** version, which is the *best ascend strategy*. In this strategy, the next tour is the *best* improving solution found in the 2-swap neighborhood as given in Algorithm 3.

7. Experimentation design

The experimentation design is as the following. Three different CPU's and three different GPU's are used to run the two different 2-opt variants on a selected number of asymmetric TSP instances (ATSP). The only measure is the time complexity.

The problem instances of the ATSP was obtained from the TSP library [34]. The following problems were selected due to differing city sizes as given in **Table 1**.

The machine specifications is given in **Table 2**. Three separate machines were used with differing CPUs and GPUs. Two machines were on a Windows 10 operating system and the other is a Central Washington University Supercomputer cluster running Ubuntu [35]. Machine 2 and 3 utilized headless GPU's.

Data	Cities
ft70	70
ftv64	65
ftv170	171
kro124p	100
rbg323	323
rbg358	358
rbg403	403

Table 1.
TSP instances and number of cities.

Specifications	Machine 1		Machine 2		Machine 3	
Processor	Intel i7-9750H	GTX 1050	Intel i7-7800X	Titan Xp	Power 8	P100
Memory	16 GB	2 GB	32 GB	12 GB	32 GB	16 GB
Cores	4	640	6	3840	6	3584
OS	Win10		Win10		Ubuntu	
Language	C++	CUDA-C	C++	CUDA-C	C++	CUDA-C
IDE	Visual Studio 17		Visual Studio 17		Makefile	
Cost (USD)		$200		$1500		$15,000

Table 2.
Machines specifications.

Algorithm 2: LS2OPT sequential version. The *Swap(T,j,i)* procedure swaps *j*−th and *i*−th cities of tour *T*

---

1 **for** *i=1* **to** N-1 **do**
2   **for** *j=i+1* **to** N **do**
3       $T \leftarrow$ Swap(*T,i,j*)
4       $f_T \leftarrow f(T)$
5   **if** $f_T < f_S$ **then**
6          $S \leftarrow T$
7          $f_S \leftarrow f_T$
8   *ImprovementFound* $\leftarrow$ True
9   break
10   **end**
11       $T \leftarrow$ Swap(*T,j,i*)
12   **end**
13 **end**
14 return S

Algorithm 3: MLS2OPT sequential version. The *Swap(T,idx,i)* procedure swaps *idx*−th and *i*−th cities of tour *T*, where *idx*−th is the best 2-swap schedule *j*−th index found after iteration

---

1 **for** *i=1* **to** N-1 **do**
2   **for** *j=i+1* **to** N **do**
3       $T \leftarrow$ Swap(*T,i,j*)
4       $f_T \leftarrow f(T)$
5   **if** $f_T < f_S$ **then**
6          $f_{S_j} \leftarrow f_T$
7   *ImprovementFound* $\leftarrow$ True
8   **end**
9   **end**
10   **if** *ImprovementFound* **then**
11       $idx \leftarrow \min(f_{S_j})$
12   **end**
13    $T \leftarrow$ Swap(*T,idx,i*)
14 **end**
15 return S

8. Results and analysis

The results are grouped by the machine architectures, as there is a dependency between the CPU and GPU. Thirty experimentations was done of each problem instance on each machine for each algorithm and the average time is given in the tables (* in msec). The *percentage relative difference* (PRD) is calculated between the *CPU* and *GPU* times as given in Eq. (3). Negatives values (given as bolded text in the tables) indicate that the GPU execution is faster.

$$PRD = ((GPU - CPU)/CPU) \cdot 100 \qquad (3)$$

The first part of the first machine experiment results of the LS2OPT and its CUDA variant is given in **Table 3**. The first column is the problem instances and the second and third column is the CPU and GPU average results of the LS2OPT in milliseconds. The final column is the PRD results. From all the results, apart from the *ftv64* instance, the GPU produced faster results. The *average* time was **22480.28** ms for the CPU and **2168.57** ms for the GPU. The average PRD was −**47.29**% for all experiments. A deeper analysis shows that for the larger instances, the PRD was over 80%.

The plot of the execution time is given in **Figure 4** where the execution speedup is clearly identifiable for the larger instances.

The second part of the first machine experimentation is the MLS2OPT and its CUDA variant and the result are given in **Table 4**. For all the problem instances, the execution time for the GPU was significantly better. The *average* time was **14183.85** ms for the CPU and **1854.28** ms for the GPU. The average PRD was −**52.55**% for all experiments. Apart from two instances, all the other were above 85% PRD.

The plot of the execution time is given in **Figure 5**, where the execution speedup is linearly identifiable for the larger instances.

The first part of the second machine experiment results of the LS2OPT and its CUDA variant is given in **Table 5**. As the NVidia Titan Xp is a dedicated headless TESLA category GPU, the computational times are better than the CPU for all the results. The *average* time was **12157.14** ms for the CPU and **857** ms for the GPU. The average PRD was −**64.92**% for all experiments. A deeper analysis shows that for the larger instances, the PRD was over 90%. As the transfer overhead for the *PCIe* bus is

Data	Intel i7-9750H LS2OPT	Nvidia GTX 1050 LS2OPTCUDA	PRD (%)
ft70	42	34	**−19.047**
ftv64	14	30	114.29
ftv170	322	87	**−72.98**
kro124p	580	111	**−80.86**
rbg323	43,854	2963	**−93.24**
rbg358	51,069	4096	**−91.98**
rbg403	61,481	7859	**−87.22**
Average	**22480.28**	**2168.57**	**−47.29**

*All results are in milliseconds (ms).

Table 3.
*Results of the experiments of **Intel i7-9750H** and **NVidia GTX 1050** on the LS2OPT and LS2OPTCUDA algorithms.*

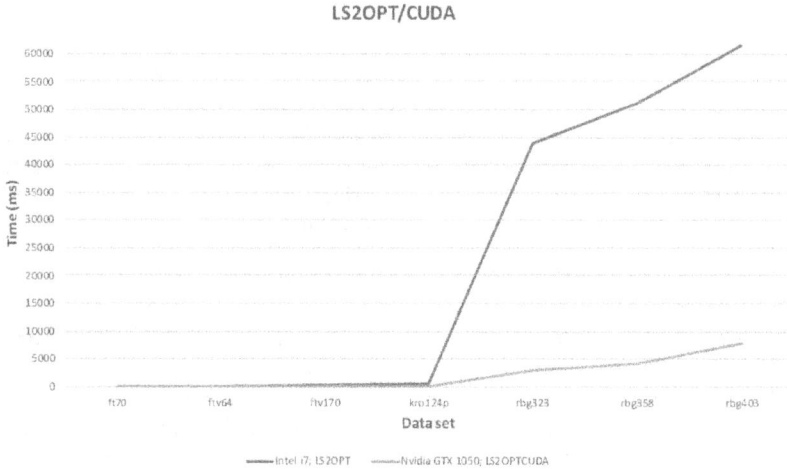

Figure 4.
*Figure for the experiments of **Intel i7** and **NVidia GTX 1050** on the LS2OPT and LS2OPTCUDA algorithms.*

Data	Intel i7-9750H MLS2OPT	Nvidia GTX 1050 MLS2OPTCUDA	PRD (%)
ft70	37	21	−43.24
ftv64	26	52	100.00
ftv170	619	78	−87.40
kro124p	303	75	−75.25
rbg323	21,205	2525	−88.09
rbg358	31,330	3775	−87.95
rbg403	45,767	6454	−85.90
Average	**14183.85**	**1854.28**	**−52.55**

All results are in milliseconds (ms).

Table 4.
*Results of the experiments of **Intel i7-9750H** and **NVidia GTX 1050** on the MLS2OPT and MLS2OPTCUDA algorithms.*

compensated by more extensive experimentation, larger instances performed faster on the GPU.

The plot of the execution time is given in **Figure 6**, where the execution speedup is clearly identifiable for the larger instances.

The second part of the second machine experimentation is the MLS2OPT and its CUDA variant and the result are given in **Table 6**. For all the problem instances, the execution time for the GPU was significantly better. The *average* time was **7955.28** ms for the CPU and **616.28** ms for the GPU. The average PRD was −**63.39**% for all experiments. The three larger instances were above 90% PRD.

The plot of the execution time is given in **Figure 7**, where the execution speedup is linearly identifiable for the larger instances.

The first part of the third machine experiment results of the LS2OPT and its CUDA variant is given in **Table 7**. Generally, the NVidia P100 is regarded as an industry leading GPU solution for scientific computing. This is coupled with the IBM Power 8 CPU Architecture. For all the problem instances the result was

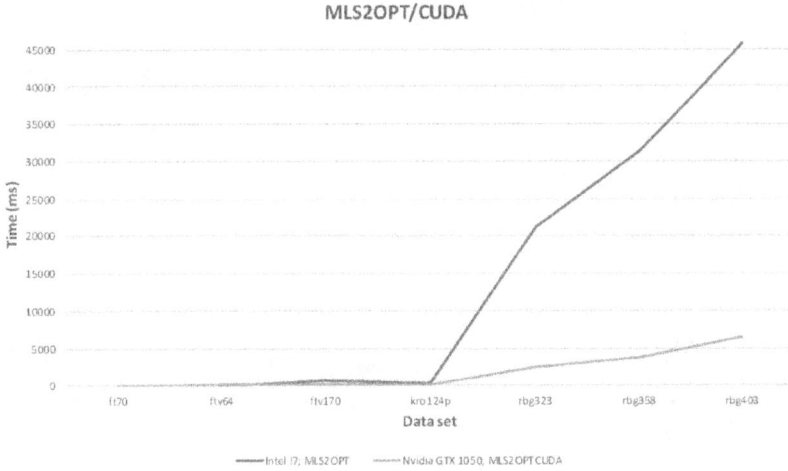

Figure 5.
*Figure for the experiments of **Intel i7** and **NVidia GTX 1050** on the MLS2OPT and MLS2OPTCUDA algorithms.*

Data	Intel i7-7800X LS2OPT	NVidia Titan Xp LS2OPTCUDA	PRD (%)
ft70	21	18	−14.29
ftv64	12	8	−33.33
ftv170	183	77	−57.92
kro124p	306	94	−69.28
rbg323	23,619	1467	−93.79
rbg358	27,614	1848	−93.31
rbg403	33,345	2487	−92.54
Average	**12157.14**	857	**−64.92**

All results are in milliseconds (ms).

Table 5.
*Results of the experiments of **Intel i7-7800X** and **NVidia titan Xp** on the LS2OPT and LS2OPTCUDA algorithms.*

significantly better. The *average* time was **28592.43** ms for the CPU and **1536.42** ms for the GPU. The average PRD was −**87.83**% for all experiments.

The plot of the execution time is given in **Figure 8**, where the execution speedup is clearly identifiable for the larger instances.

The second part of the third machine experimentation is the MLS2OPT and its CUDA variant and the result are given in **Table 8**. For all the problem instances, the execution time for the GPU was again significantly better. The *average* time was **23429.14** ms for the CPU and **751** ms for the GPU. The average PRD was −**92.78**% for all experiments. The PRD is the highest of all experiments.

The plot of the execution time is given in **Figure 9**, where the execution speedup is linearly identifiable for the larger instances.

The final comparison is of the three GPU's on the two separate algorithms. **Figure 10** shows the values of the three GPU's on the problem instances for the *LS2OPTCUDA* algorithm. For the small sized problem, the timing is not significantly distinct. The distinction only becomes variable when the instance sizes increase. Overall, the NVidia Titan Xp is the best performing GPU for this algorithm.

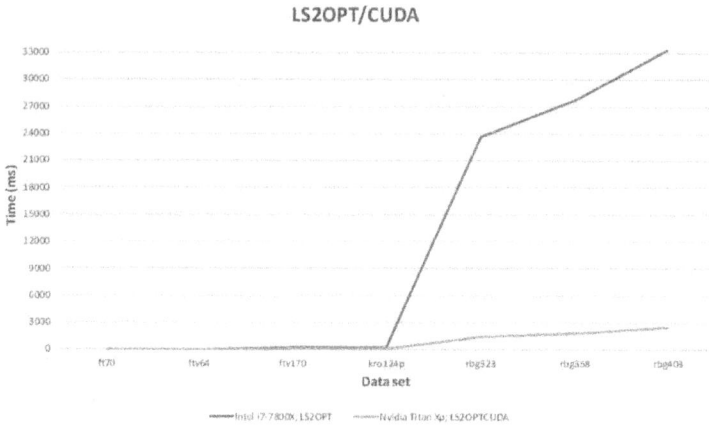

Figure 6.
*Figure for the experiments of **Intel i7-7800X** and **NVidia titan Xp** on the LS2OPT and LS2OPTCUDA algorithms.*

Data	Intel i7-7800X MLS2OPT	NVidia Titan Xp MLS2OPTCUDA	PRD (%)
ft70	20	16	−20.00
ftv64	11	8	−27.27
ftv170	321	58	−81.93
kro124p	164	56	−65.85
rbg323	11,517	941	−91.83
rbg358	17,109	1059	−93.81
rbg403	24,445	2176	−91.10
Average	**7955.28**	**616.28**	**−63.39**

All results are in milliseconds (ms).

Table 6.
*Results of the experiments of **Intel i7-7800X** and **NVidia titan Xp** on the MLS2OPT and MLS2OPTCUDA algorithms.*

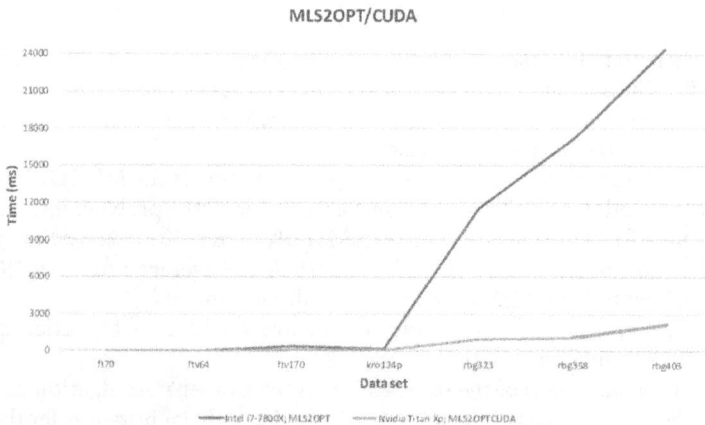

Figure 7.
*Figure for the experiments of **Intel i7-7800X** and **NVidia titan Xp** on the MLS2OPT and MLS2OPTCUDA algorithms.*

Data	Power 8 LS2OPT	NVidia P100 LS2OPTCUDA	PRD (%)
ft70	57	10	−82.46
ftv64	23	6	−73.91
ftv170	430	75	−82.56
kro124p	754	61	−91.91
rbg323	61,775	3245	−94.75
rbg358	64,419	3587	−94.43
rbg403	72,689	3771	−94.81
Average	**28592.43**	**1536.42**	**−87.83**

*All results are in milliseconds (ms).

Table 7.
*Results of the experiments of **power 8** and **NVidia P100** on the LS2OPT and LS2OPTCUDA algorithms.*

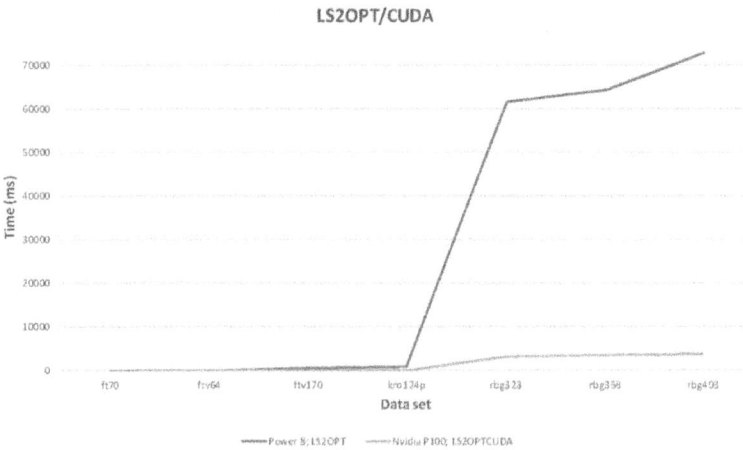

Figure 8.
*Figure for the experiments of **power 8** and **NVidia P100** on the LS2OPT and LS2OPTCUDA algorithms.*

Data	Power 8 MLS2OPT	NVidia P100 MLS2OPTCUDA	PRD (%)
ft70	53	7	−86.79
ftv64	33	4	−87.88
ftv170	811	52	−93.59
kro124p	385	35	−90.91
rbg323	30,120	1124	−96.27
rbg358	44,709	1215	−97.28
rbg403	87,893	2820	−96.79
Average	**23429.14**	**751**	**−92.78**

*All results are in milliseconds (ms).

Table 8.
*Results of the experiments of **power 8** and **NVidia P100** on the MLS2OPT and MLS2OPTCUDA algorithms.*

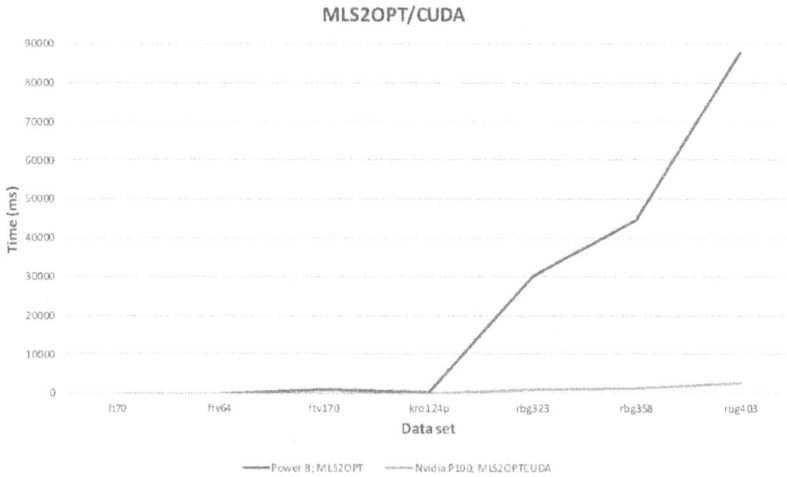

Figure 9.
*Figure for the experiments of **power 8** and **NVidia P100** on the MLS2OPT and MLS2OPTCUDA algorithms.*

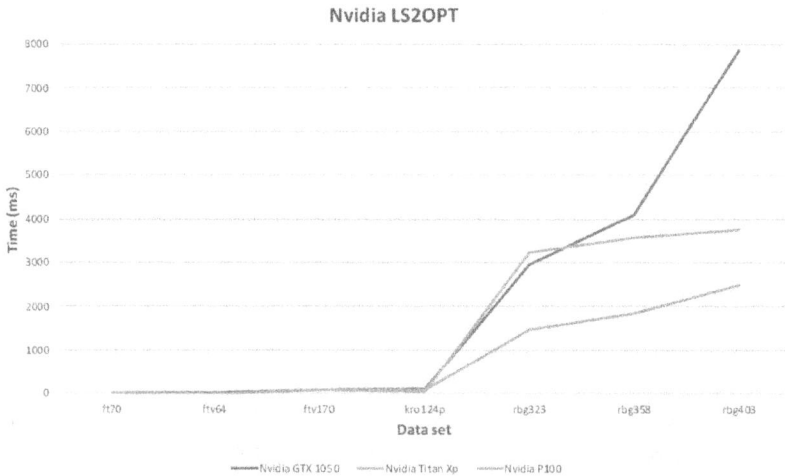

Figure 10.
Figure for the experiments of the three NVidia GPU's for the LS2OPTCUDA algorithm.

Figure 11 shows the results of the *MLS2OPTCUDA* algorithm on the problem. As with the previous case, the distinction only becomes obvious for large sized problem instances. Again the NVidia Titan Xp is the best performing GPU for this algorithm.

9. Algorithm comparison

This section discusses the tour cost obtained by the two different 2-OPT approaches developed here compared with published research. The first comparison is done with the best known solution in literature, which can be obtained from the TSPLib [36].

Table 9 gives the comparison results between the optimal and the results obtained from the LS2OPTCUDA and MLS2OPTCUDA algorithms on the P100

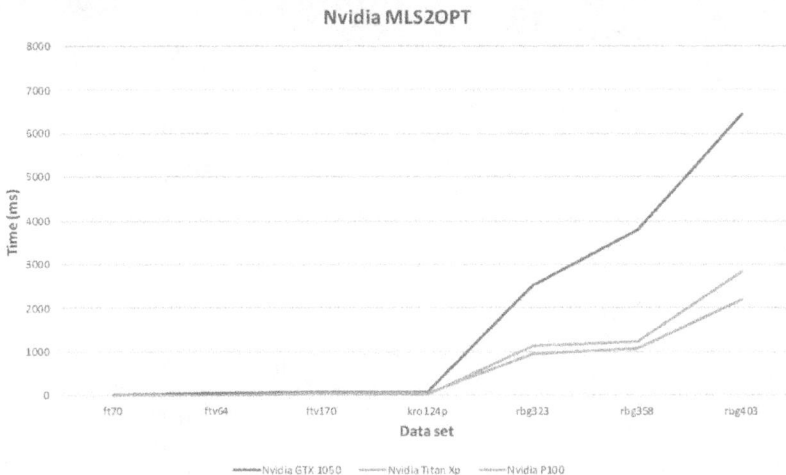

Figure 11.
Figure for the experiments of the three NVidia GPU's for the MLS2OPTCUDA algorithm.

Data	Optimal	LS2OPT	PRD (%)	MLS2OPT	PRD (%)
ft70	38,673	43,163	−10.40	43,310	−10.71
ftv64	1839	2744	−32.98	2554	−28.00
ftv170	2755	4559	−39.57	4510	−38.91
kro124p	36,230	58,014	−37.55	55,011	−34.14
rbg323	1326	1681	−21.12	1535	−13.62
rbg358	1163	1625	−28.43	1459	−20.29
rbg403	2465	2710	−9.04	2598	−5.12
Average	12064.43	16356.57	−25.58	**15853.86**	−21.54

Table 9.
Comparison of 2OPT vs. optimal values.

GPU. The results are compared using the PRD Eq. (3). The GPU is replaced with the optimal value and the CPU is replaced by the obtained result.

The PRD values comparison shows that the LS2OPT is at most 40% away from the optimal value for *ftv170* instance and − 9% for the *rbg403* instance. For the MLS2OPT comparison, the PRD is −39% from the optimal value for *ftv170* instance and − 5% for the *rbg403* instance. On average, the MLS2OPT is a better performing algorithm with an average of 15853.86 against 16356.57 for the LS2OPT algorithm. A plot of the comparison values is given in **Figure 12**.

The second comparison is now done with four different evolutionary algorithms as given in **Table 10**. Theses are the Discrete Particle Swarm Optimization (DPSO) algorithm [37], Discrete Self-Organizing Algorithm (DSOMA) [38], Enhanced Differential Evolution (EDE) algorithm and the Chaos driven Enhanced Differential Evolution (EDE_C) algorithm [17]. The DPSO and DSOMA algorithms were revised for the TSP problem and the 2-OPT local search was removed from the algorithms to compare the results without any local search implemented. EDE and EDE_C are published algorithms however only three instances were published. Both these algorithms had the 2-OPT local search embedded in them.

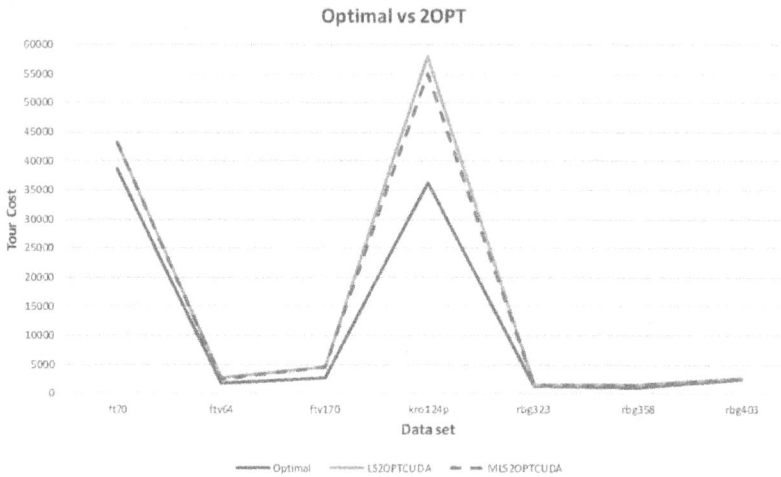

Figure 12.
Figure for the comparison of 2-OPT against global optimal values [36].

Data	MLS2OPT	DPSO	DSOMA	EDE	EDE$_C$
ft70	43,310	54,444	51,325	40,285	**39,841**
ftv64	**2554**	4711	4423	—	—
ftv170	4510	19,102	9522	6902	**4578**
kro124p	55,011	113,153	75,373	41,180	**39,574**
rbg323	**1535**	4852	4523	—	—
rbg358	**1459**	5692	4874	—	—
rbg403	**2598**	6373	4427	—	—
Average	**15853.86**	29761.00	22066.71	—	—

Table 10.
MLS2OPT vs. evolutionary algorithms.

From the results, it was obvious that evolutionary algorithms without local search heuristics are not as effective as the *2-opt local search* heuristic or algorithms with both *directed* and *local* search combined. Therefore, it is important to combine these two algorithms as in [39]. As reported in [39] that the execution time of local search can be around 95–99% of the total run time of the algorithm, it is viable to accelerate the local search heuristics.

10. Conclusions

This chapter introduces a CUDA accelerated 2-opt algorithm for the TSP problem. As one of the most common and widely used approaches to solve the problem, the 2-opt approach can be considered as canonical in the field.

GPU programming, especially CUDA has gained significant traction for high performance computing. Readily available hardware has made programming a much easier and available task.

Two variants of the 2-opt algorithm have been coded in CUDA to show the acceleration of computational time. This has been tested against a sample of test instances from literature. From the results obtained, it is clear that even for a relatively cheap GPU such as the GTX 1050 the performance improvement is significant, especially for larger sized problem instances. These were compared against industry leading CPU's such as Intel i7-X series and IBM Power 8.

One of the interesting aspects was that the Titan Xp performed better than the P100 for these instances. It is difficult to identify the reasons, as the same code was deployed on all machines, however the IBM and Intel architecture differences and different C/C++ compiler usage may have affected the performance. The physical configuration of the GPU's inside the hardware and its connection to the motherboard and memory bandwidth issues could also add to the time overhead. However, when analyzing the cost-performance of the GPU's then the $1500 Titan Xp is a better GPU than the $15,000 P100 in this case.

However, the clear distinction is that there is a significant improvement to be had when applying the CUDA version of the 2-opt algorithm. The next direction of this research is to combine it with powerful swarm meta-heuristics with a layered approach, and try and solve very large TSP instances.

Author details

Donald Davendra[1][*][†], Magdalena Metlicka[2][†] and Magdalena Bialic-Davendra[1][†]

1 Central Washington University, Ellensburg, USA

2 Honeywell Engineering Aerospace, Brno, Czech Republic

*Address all correspondence to: donald.davendra@cwu.edu

† These authors contributed equally.

IntechOpen

References

[1] Englert M, Roglin H, Vocking B. Worst case and probabilistic analysis of the 2-opt algorithm for the TSP. Algorithmica. 2014;**68**:190-264

[2] Van Leeuwen J, Schoon A. Untangling a traveling salesman tour in the plane. In: Proceedings of the 7th International Workshop on Graph-Thoeratical Concepts in Computer Science. The Netherlands: Rijksuniversiteit. Vakgroep Informatica; 1981. pp. 87-98

[3] Johnson D, McGeoch L. The traveling salesman problem: A case study in local optimization. In: Aarts E, Lenstra J editors, Local Search in Combinatorial Optimization. Hoboken, NJ, USA: John Wiley and Sons; 1997

[4] Farber R. CUDA Application Design and Development. Burlington, MA, USA: Morgan Kaufmann; 2012

[5] Lawler EL, Lenstra JK, Kan AR, Shmoys DB. The Traveling Salesman Problem: A Guided Tour of Combinatorial Optimization. Vol. 3. New York: Wiley; 1985

[6] Davendra D, editor. Traveling Salesman Problem, Theory and Applications. Rijeka: IntechOpen; 2010

[7] Laporte G. The traveling salesman problem: An overview of exact and approximate algorithms. European Journal of Operations Research. 1992; **59**(2):231-247

[8] Li H, Alidaee B. Tabu search for solving the black-and-white travelling salesman problem. Journal of the Operational Research Society. 2016; **67**(8):1061-1079

[9] Kirkpatrick S, Gellat C, Vecchi M. Optimization by simulated annealing. Science. 1983;**220**(4598):671-680

[10] Grefenstette J, Gopal R, Rosmaita B, Van Gucht D. Genetic algorithms for the traveling salesman problem. In: Proceedings of the First International Conference on Genetic Algorithms and their Applications. New Jersey: Lawrence Erlbaum; 1985. pp. 160-168

[11] Oliver I, Smith D, Holland JR. Study of permutation crossover operators on the traveling salesman problem. In: Genetic Algorithms and their Applications: Proceedings of the Second International Conference on Genetic Algorithms; July 28–31, 1987 at the Massachusetts Institute of Technology, Cambridge, MA, Hillsdale, NJ, L. Erlhaum Associates; 1987

[12] Yu B, Yang Z-Z, Yao B. An improved ant colony optimization for vehicle routing problem. European Journal of Operations Research. 2009; **196**(1):171-176

[13] Tang K, Li Z, Luo L, Liu B. Multi-strategy adaptive particle swarm optimization for numerical optimization. Engineering Applications of Artificial Intelligence. 2015;**37**:9-19

[14] Yang X-S, Deb S. Cuckoo search via levy flights. In: World Congress on Nature & Biologically Inspired Computing, 2009. NaBIC 2009. NY, USA: IEEE Publications; 2009. pp. 210-214

[15] Yang X-S. Firefly algorithms for multimodal optimization. In: Stochastic Algorithms: Foundations and Applications. Berlin, Heidelberg, Germany: Springer; 2009. pp. 169-178

[16] Osabaa E, Sera D, Sadollahd A, Miren Nekane Bilbaob J, Camachoe D. A discrete water cycle algorithm for solving the symmetric and asymmetric traveling salesman problem. Applied Soft Computing. 2018;**71**: 277-290

[17] Davendra D, Zelinka I, Senkerik R, Bialic-Davendra M. Chaos driven evolutionary algorithm for the traveling salesman problem. In: Davendra D, editor. Traveling salesman problem. Rijeka: IntechOpen. DOI: 10.5772/13107

[18] Li L, Cheng Y, Tan Y, Niu B. A discrete artificial bee colony algorithm for TSP problem. In: Proceedings of the 7th International Conference on Intelligent Computing: Bio-Inspired Computing and Applications (ICIC'11). Berlin, Heidelberg: Springer-Verlag; 2011. pp. 566-573. DOI: 10.1007/978-3-642-24553-4_75

[19] Davendra D, Zelinka I, Pluhacek M, Senkerik R. DSOMA—Discrete self-organising migrating algorithm. In: Self-Organizing Migrating Algorithm: Methodology and Implementation. Berlin, Heidelberg, Germany: Springer; 2016. pp. 51-63

[20] Croes G. A method for solving traveling-salesman problems. Operations Research. 1958;**6**(6):791-812

[21] Shen L. Computer solutions of the traveling salesman problem. Bell System Technical Journal. 1965;**44**(10): 2245-2269

[22] Savelsbergh M. An efficient implementation of local search algorithms for constrained routing problems. European Journal of Operational Research. 1990;**47**(1):75-85

[23] Johnson D, McGeoch L. The traveling salesman problem: A case study in local optimization. Local search in combinatorial optimization. 1997;**1**: 215-310

[24] Gutin G, Punnen A. The Traveling Salesman Problem and Its Variations. Vol. 12. Berlin, Heidelberg, Germany: Springer; 2002

[25] Kim B, Shim J, Zhang M. Comparison of tsp algorithms. In:

Project for Facilities Planning and Materials Handling. 1998

[26] Kronos Group. Available from: https://www.khronos.org/ [Accessed: 24 May 2020]

[27] Sanders J, Kandrot E. CUDA by Example. 1st Print Edition. Addison-Wesley; 2010

[28] Kirk D, Wen-mei W. Programming Massively Parallel Processors: A Hands-on Approach. Newnes; 2012

[29] NVIDIA: Cuda C Programming Guide. Santa Clara, CA, USA: NVIDIA Corporation; 2020

[30] NVIDIA: Kepler gk110. Santa Clara, CA, USA: NVIDIA Corporation; 2012

[31] Metlicka M. Framework for scheduling problems [master thesis]. Czech Republic: Technical University of Ostrava; 2015

[32] Metlicka M, Davendra D, Hermann F, Meier M, Amann M. GPU accelerated NEH algorithm. In: 2014 IEEE Symposium on Computational Intelligence in Production and Logistics Systems (CIPLS), Orlando, FL; 2014. pp. 114-119. DOI: 10.1109/CIPLS.2014.7007169

[33] NVIDIA. Cuda C Best Practices Guide [Online]. 2020

[34] TSP Library ATSP Dataset. Available from: http://elib.zib.de/pub/mp-testdata/tsp/tsplib/atsp/index.html [Accessed: 02 January 2020]

[35] CWU Turing Supercomputer. Available from: http://www.cwu.edu/faculty/turing-cwu-supercomputer [Accessed: 20 February 2020]

[36] TSP Library ATSP Best Known Solutions. Available from: http://elib.zib.de/pub/mp-testdata/tsp/tsplib/atsp-sol.html [Accessed: 17 May 2020]

[37] Wang X, Tang L. A discrete particle swarm optimization algorithm with self-adaptive diversity control for the permutation flowshop problem with blocking. Applied Soft Computing. 2012;**12**:652-662

[38] Davendra D, Bialic-Davendra M. Discrete self organizing algorithm for pollution vehicle routing problem. In: Proceedings of the genetic and evolutionary computation conference 2020 (GECCO 20 companion). New York, NY, USA: ACM; 2020. p. 8. DOI: 10.1145/3377929.3398076

[39] Merz P, Freisleben B. Genetic local search for the TSP: New results. In: Proceedings of 1997 IEEE International Conference on Evolutionary Computation (ICEC '97), Indianapolis, IN, USA; 1997. pp. 159-164. DOI: 10.1109/ICEC.1997.592288

Solution Attractor of Local Search System: A Method to Reduce Computational Complexity of the Traveling Salesman Problem

Weiqi Li

Abstract

The traveling salesman problem (TSP) is presumably difficult to solve exactly using local search algorithms. It can be exactly solved by only one algorithm—the enumerative search algorithm. However, the scanning of all possible solutions requires exponential computing time. Do we need exploring all the possibilities to find the optimal solution? How can we narrow down the search space effectively and efficiently for an exhausted search? This chapter attempts to answer these questions. A local search algorithm is a discrete dynamical system, in which a search trajectory searches a part of the solution space and stops at a locally optimal point. A solution attractor of a local search system for the TSP is defined as a subset of the solution space that contains all locally optimal tours. The solution attractor concept gives us great insight into the computational complexity of the TSP. If we know where the solution attractor is located in the solution space, we simply completely search the solution attractor, rather than the entire solution space, to find the globally optimal tour. This chapter describes the solution attractor of local search system for the TSP and then presents a novel search system—the attractor-based search system—that can solve the TSP much efficiently with global optimality guarantee.

Keywords: local search, global optimization, computational complexity, dynamical system, combinatorial optimization, solution attractor

1. Introduction

What it is that makes the TSP difficulty? The difficulty of the TSP is associated with the combinatorial explosion of potential solutions in the solution space. When a TSP instance is large, the number of possible tours in the solution space is so large as to forbid an exhausted search for the optimal tour. Numerous approaches to solving the TSP have been published. Some algorithms such as enumerative search, branch-and-bound search, and linear programming are exact approaches but lack efficiency. Other approximate algorithms, based on heuristics, are quick to find a good tour but lack effectiveness and robustness. Modern approximate algorithms, with today's fast computers, can find good solutions for extremely large TSP instances within a reasonable time, which are with a high probability just 2–3% away from the optimal tour [1–3].

Most approximate algorithms have been based on or derived from a general search technique known as *local search*. Local search algorithms iteratively explore the neighborhoods of solutions trying to improve the current solution by local changes. However, the scope of a single search trajectory is limited by the neighborhood definition. Both the TSP and local search have been hot research topics for decades, and many aspects of them have been studied. However, there is still a variety of open questions. The study of local search for the TSP continues to be a vibrant, exciting, and fruitful endeavor in combinatorial optimization, computational mathematics, and computer science.

A local search algorithm is essentially in the domain of dynamical systems. The goal of a dynamical system analysis is to capture the distinctive properties of certain points in the state space for a given dynamical system. The attractor theory of dynamical systems is a natural paradigm that provides the necessary and sufficient theoretical foundation to study the convergent behavior of a local search system. The TSP is believed to be NP-hard because we do not have an efficient enumerative search system for the problem. Do we need to examine all possibilities in order to solve the problem? Can we quickly narrow down the search space to a small region in which the optimal solution is located and then search that small region completely to find the optimal solution? This chapter attempts to use the solution attractor concept to answer these questions. If we can quickly identify that small region, the solution attractor, and then search that region thoroughly in reasonable time, the computational complexity of the problem can be dramatically reduced or may not exist. This chapter introduces the solution attractor concept, which not only helps us understand the behavior of a local search system for the TSP but also offers an important method to solve the problem efficiently with global optimality guarantee. This chapter presents a novel search algorithm—the attractor-based search system (ABSS)—that is a simple and quick global search system for the TSP.

2. Reframing the TSP definition

A problem is the frame into which the solutions fall. By changing the frame, we can change the range of possible solutions and scope of the optimal solutions. The classic TSP is defined as a complete graph $Q = (V, E, C)$, where $V = \{v_i : i = 1, 2, \ldots, n\}$ is a set of n nodes, $E = \{e(i,j) : i, j = 1, 2, \ldots, n; i \neq j\}$ is an $n \times n$ edge matrix containing the set of edges that completely connects the n nodes, and $C = \{c(i,j) : i, j = 1, 2, \ldots, n; i \neq j\}$ is an $n \times n$ cost matrix holding a set of costs between nodes. A tour $s \in S$ is a closed tour that visits every node exactly once and returns to the starting node at the end. The solution space S contains a finite set of all feasible tours. The goal of the TSP is to find a tour s^* with minimal cost:

$$s^* = \min_{s \in S} f(s) \tag{1}$$

Obviously, this definition requires a search algorithm to find any single optimal tour in the solution space for a given instance. However, many real-world optimization problems are inherently multimodal. They may contain multiple optimal solutions in their solution spaces. Finding all optimal solutions is the essential requirement for global optimization. In practice, knowledge of multiple optimal solutions is essentially helpful, providing the decision-maker with multiple best options. We assume that a TSP instance contains h $(h \geq 1)$ optimal tours in the solution space S and denotes S^* as the set of h optimal tours. Under global optimization frame, the objective of the TSP is to find the set of optimal tours $S^* \subset S$:

$$S^* = \arg \left[\min_{s \in S} f(s) \right] = [s_1^*, s_2^*, ..., s_h^*] \tag{2}$$

For a given TSP instance, we do not know the number of optimal tours in the solution space until we find all of them. Obviously, this reframed TSP definition becomes even more difficult to solve. To solve this reframed TSP, we need a search algorithm that converges not just in value but also in solution. *Convergence in value* means that a search system can find any one of the optimal solutions in the solution space eventually. *Convergence in solution* means that the search system can identify the same set of optimal solutions in the solution space over and over again.

Usually, the edge matrix E is not necessary to be included in the TSP definition because the TSP is a complete graph. However, the matrix E is a powerful data structure that can shift our point of view so that we can uncover alternative approaches. One factor contributing to algorithmic difficulty is that we lack a data structure that links the structure of the problem and the behavior of the search algorithm and that can make the complex search space traceable and tractable. It may be unreasonable to expect a search algorithm to be able to solve any problem without taking into account the structure and properties of the problem. Local search algorithms may not require much problem-specific knowledge in order to generate good solutions. However, in order to solve a problem exactly, we should design a search algorithm that is based on the structure of the problem at hand.

3. Solution attractor of local search system for TSP

A dynamical system is a model to describing the temporal evolution of a system in its state space [4–9]. The theory of dynamical system is an extremely broad area of study. The study of dynamical systems has discovered that many dynamical systems exhibit attracting behavior in the system trajectories. In such a system, all initial states tend to evolve toward a single final state or a set of final states. This single state or a set of states is called *attractor*. A heuristic local search system essentially is a discrete dynamical system and therefore natural in the domain of dynamical systems.

A local search system has a solution space S, a set of times T (iterations of search), and a search function $f : S \times T \to S$ for temporal evolution that gives the consequent to a solution $s \in S$. A search trajectory is the sequence of solutions of a local search system at successive time steps in the form $s(t + 1) = f(s(t))$. The behavior of a search trajectory can be understood as a process of iterating a function $f(s)$. Questions about the behavior of a local search system over time are actually the questions about its search trajectories. Let us denote s_0 as an initial point of a search trajectory, f^t as the fth iterate of the function $f(s)$, and a locally optimal solution s' as the limit of the convergent search trajectory $s_0, f(s_0), f^2(s_0), ..., f^t(s_0), ...$; then

$$f(s') = f\left(\lim_{t \to \infty} f^t(s_0) \right) = \lim_{t \to \infty} f^{t+1}(s_0) = s' \tag{3}$$

For the TSP, a search trajectory leads to a sequence of tours $s_0, s_1, s_2, ..., s_t$, where s_0 is an arbitrary initial tour and s_t is the final tour at the end of search after t iterations. This time series represents a part of the solution space searched by this search trajectory. The globally optimal tour s^* is the target point of a search trajectory. Due to the constraint of the neighborhood search structure, a search trajectory rarely reaches the target point and eventually stops at a locally optimal tour s'. In a heuristic local search system, different initial points and randomness in the search process lead to a complex search behavior and generate different search trajectories. There are no two search trajectories that are exactly alike. Different search trajectories explore different regions of the solution space and stop at different final points. Since all search trajectories have the same target point, they move toward the same direction and finally stop in the same target region in the solution space. This target region is called the *solution attractor*, denoted as A. Roughly speaking, the solution attractor of a local search system is a closed region of the solution space toward which a search trajectory tends to evolve regardless of the starting point. A solution attractor is the equilibrium level of the system dynamics. At this level, all search trajectories will stop moving, and therefore the solution attractor consists of all locally optimal tours. A single search trajectory typically converges to either one of the points in the solution attractor. Since the globally optimal tour is a special case of locally optimal tours, it is undoubtedly embodied in the solution attractor, that is, $s^* \in A$ and $A \subset S$. **Figure 1** summaries the concepts of search trajectories and solution attractor in a local search system. Illustrating search trajectories and solution attractor of a local search system as 2-D object is a valid metaphor for understanding how a local search system might proceed. The solution attractor A of a local search system has the following properties [10–12]:

- Invariance, i.e., $\forall s' \in A, f^t(s') = s'$ and $f^t(A) = A$ for all time t.

- Attractiveness, i.e., $\forall s_i \in S, f^t(s_i) \in A$ for sufficient time t.

- Convexity, i.e., all locally optimal tours in A are gathered in an extremely small region of the solution space.

- Centrality, i.e., the best of these locally optimal tours (the globally optimal tour) is located centrally with respect to the other locally optimal tours.

- Irreducibility, i.e., the solution attractor A contains a limit number of invariant locally optimal tours.

In general term, for a TSP instance with h $(h \geq 1)$ optimal tours, the local search system will have h solution attractors $(A_1, A_2, \dots, A_h)$ that attract all search trajectories. Each of the solution attractors has its own set of locally optimal tours,

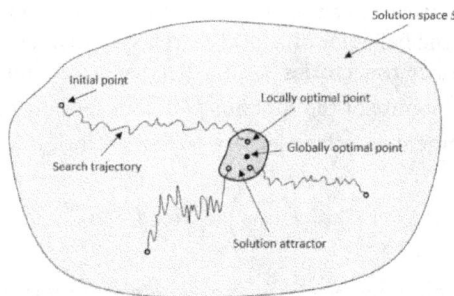

Figure 1.
Search trajectories and solution attractor in a local search system.

surrounding a globally optimal tour s_i^* $(i = 1, 2, ... , h)$. The search trajectories will explore many different regions of the solution space and converge to these solution attractors. A particular search trajectory will converge into one of these solution attractors. The set of locally optimal tours generated by all search trajectories will be distributed to these h solution attractors. According to dynamical systems theory [9], the closure of an arbitrary union of attractors is still an attractor, that is, the attractor of a local search system for a multimodal TSP is a complete collection of solution attractors $A = A_1 \cup A_2 \cup ... \cup A_h$ and $A \subset S$.

4. The attractor-based search system for TSP

Figure 2 presents the attractor-based search system (ABSS) for the TSP. In this algorithm, Q is a given TSP instance. K is the number of search trajectories used to generate K locally optimal tours. E is the edge matrix used to store the K locally optimal tours. s_i is an initial tour generated by the function Initial_Tour(), which can use any technique to construct the initial tour. s_j is a locally optimal tour generated by the function Local_Search(), which can use any local search technique. The function Update() updates the edge matrix E by recording the edge configuration of tour s_j into E. Finally, the function Exhausted_Search() searches the matrix E completely using any enumerative search technique and outputs the set of the found globally optimal tours, S^*. The search strategy behind the ABSS is simple and effective: we first identify the small regions—the solution attractors—in which the globally optimal tours are located, and then we search these small regions completely to find the globally optimal tours. In this strategy, we avoid searching the large unnecessary region of the solution space so that the search time is dramatically reduced. The ABSS shows strong features of effectiveness, flexibility, adaptability, and scalability. It can be implemented in many different ways: serial or parallel. The computational model in ABSS is inherently parallel and can support the exploitation of massive parallelism. If the ABSS is implemented using proper number of concurrent processors, it can deal with dynamic TSP in real time:

The critical element in the ABSS is the edge matrix E. Few search algorithms have used the edge matrix E in their search processes. An edge is the most basic element in a tour. It is a connection between two nodes and contains pieces of information about $(n - 2)!$ tours that go through it. A tour is a list of ordered nodes and has an edge configuration in the matrix E, as an example illustrated in **Figure 3**. Each edge has an implicit probability to be selected by a locally optimal tour. The edges in the matrix E can be divided into three groups: G-edges, globally superior

```
1   ABSS_Algorithm(Q)
2   begin
3       initialize E;
4       NumberOfTrajectories = 0;
5   repeat
6           s_i = Initial_Tour();
7           s_j = Local_Search(s_i);
8           E = Update(E, s_j);
9           NumberOfTrajectories = NumberOfTrajectories + 1;
10  until NumberOfTrajectories = K
11  S* = Exhausted_Search(E);
12  end
```

Figure 2.
The ABSS algorithm for the TSP.

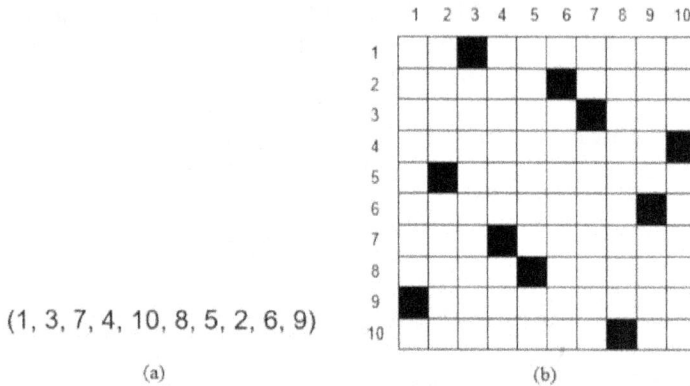

$$(1, 3, 7, 4, 10, 8, 5, 2, 6, 9)$$

(a)   (b)

Figure 3.
(a) Shows a 10-node tour and (b) shows its edge configuration in the matrix E.

edges, and bad edges. The edges that are contained in a globally optimal tour are
G-edges. A globally superior edge is the edge that is hit by many locally optimal
tours. Although each of these locally optimal tour selects this edge based on its own
neighborhood function and search path, the edge is globally superior because it is
selected by these tours from different search paths that go through different regions
of the solution space. Bad edges are the edges that are eventually discarded by all
search trajectories or selected by only few locally optimal tours. The edge configuration of a locally optimal tour consists of some G-edges, some globally superior
edges and a few bad edges. Therefore, the edge matrix E is an exploitable data
structure that plays the following roles in the ABSS:

- It is a natural data structure that can store the edge configurations of search
  trajectories and thus can visually demonstrate the asymptotic behavior of the
  search trajectories during the search. When the search trajectories reach their
  final points, it records the frequency of occurrence of each of the edges in the
  locally optimal tours.

- It is an instrument that can alter the state of what we measure for the TSP. We
  can change a tour-search process into an edge-search process, and thus the
  problem of finding the optimal tour is converted into the problem of finding a
  set of edges. The edge space represented by the edge matrix E is much simpler
  and smaller than the solution space represented by the tours.

- It is a mechanism that can transform non-deterministic local search to
  deterministic global search. Through the matrix E, we can see that the search
  trajectories actually perform the process of edge inclusion and exclusion, and the
  temporal evolution of the edge configuration matrix E generated by different
  sets of K search trajectories always converges to the same small set of edges.

A search trajectory changes its edge configuration during the search process. Let
W be the total number of edges in the matrix E, $\alpha(t)$ the number of the common
edges that are hit by all search trajectories at time t, $\beta(t)$ the number of the edges
that are hit by one or some of the search trajectories, and $\gamma(t)$ the number of the
edges that have no hit from the search trajectories. Then at any time t, we have

$$W = \alpha(t) + \beta(t) + \gamma(t) \tag{4}$$

For a given TSP instance, W is a constant value $n(n-1)/2$ for a symmetric instance or $n(n-1)$ for an asymmetric instance. We can expect that, as local search process continues, the values for both $\alpha(t)$ and $\gamma(t)$ will increase and value for $\beta(t)$ will decrease. Our experiments confirmed this inference about $\alpha(t)$, $\beta(t)$, and $\gamma(t)$. **Figure 4** illustrates the curve patterns of $\alpha(t)$, $\beta(t)$, and $\gamma(t)$. These curves cannot increase or decrease forever, and they approach to constant values as the search time continues, that is,

$$W = \lim_{t\to\infty} \alpha(t) + \lim_{t\to\infty} \beta(t) + \lim_{t\to\infty} \gamma(t) = A + B + \Gamma \qquad (5)$$

This indicates that at certain point of time, the union of the edge configurations of the search trajectories will become fixed. This aggregate edge configuration will be the edge configuration of the solution attractor at limit.

When the matrix E records the edge configurations of K locally optimal tours, the edges are partitioned into two sets: the edges with hit (hit edges) and the edges without hit (non-hit edges). The hit edges include all globally superior edges, all G-edges, and some bad edges. **Figure 5** shows the composition of edges in the matrix E after the edge configurations of K locally optimal tours are stored in it. The local search process can quickly make large number of edges become the non-hit edges. In our experiments, we found that the ration $\gamma(t)/W$ exceeds 75% easily with short search time for the symmetric TSP. This fact indicates that the edge configuration of the solution attractor contains very small percentage of the edges. Therefore, compared to the full solution space, the solution attractor is extremely small.

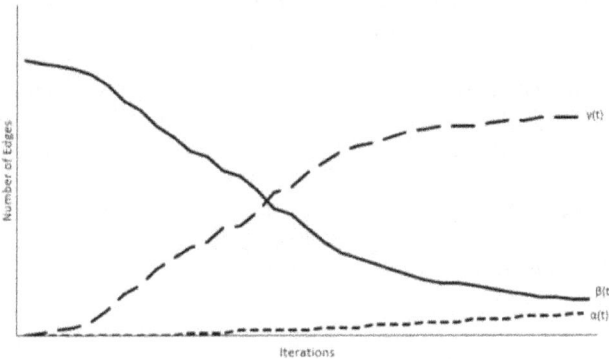

Figure 4.
The α(t), β(t), and γ(t) curves with search iterations.

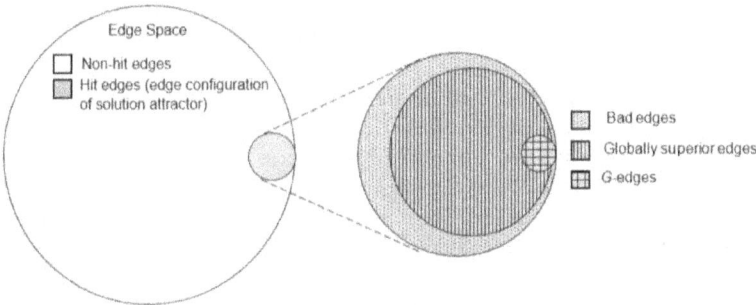

Figure 5.
The composition of edges in the matrix E after the edge configurations of K locally optimal tours are stored.

Different sets of K search trajectories will generate a little different edge configuration in the matrix E. However, the underlying edge configuration of the solution attractor in the matrix E is structurally stable because small differences in the final edge configurations generated by different sets of K search trajectories do not mean the qualitative difference in the dynamical behavior of search trajectories. The core structure of the edge configuration of the solution attractor keeps unchanged. In our experiments, we observed that in the aggregated edge configurations of the different sets of K locally optimal tours, the set of globally superior edges and the G-edges is always the same. This empirical fact indicates that a local search system actually is a deterministic system. Although a single search trajectory appears stochastic, there is an important aspect of order hidden in the local search system that makes all different sets of K search trajectories converge to the same set of core edges.

5. Global optimization and computational complexity of ABSS

In order to make sure that the ABSS is an effective and efficient search system, we should answer the following fundamental questions:

1. "How can we construct the edge configuration of the solution attractor without large number of search trajectories?" that is, "What is a proper size of K?"

2. What is the relationship between the size of the constructed solution attractor and the size of the TSP instance?

3. How does the ABSS meet the requirements of a global optimization system?

4. Is the best tour in the solution attractor the best tour in the solution space?

It is easy to verify that the edge configuration of a true solution attractor can be obtained if all search trajectories are performed and all search trajectories reach their real locally optimal points. In other words, the probability of finding all globally optimal points is one if all possible search trajectories are performed. However, the required search effort may be very huge—equivalent to enumerating all possibilities in the solution space. In fact, we can construct the edge configuration of the solution attractor with a limited number of K locally optimal tours. In a heuristic local search system, K search trajectories start a sample of initial points from a uniform distribution over the solution space S and generate a sample of locally optimal points uniformly distributed over the solution attractor A. The fundamental theory behind using K search trajectories is the information theory. According to the information theory [13], each solution point in the solution space contains some information about its neighboring points that can be modeled as mapping $\Omega_{s_i} : s_i \rightarrow \Re$, called *information* or *influence* function, which is a decreasing function of the spatial distance to the solution point s_i in the solution space. The information function value of s_i is maximum at the point and decreases gradually with the distance from that point. The notion of influence function has been used extensively in data mining, data clustering, and pattern recognition. In a local search system for the TSP, as one search trajectory is approaching to a locally optimal tour, it shares more and more edges with other search trajectories and thus collects more and more information about the other locally optimal tours and the globally optimal tour. When K search trajectories reach their end points and record their edge configurations in the matrix E, the aggregate edge configuration in the matrix E is not just a countable union of the edge configurations of the K locally optimal tours but also includes the edge configurations of all other locally optimal tours. The essential motivation behind using the edge matrix E is that a collection of K locally optimal tours is able to provide whole information about all locally optimal tours and the matrix E is a tool that put all pieces of puzzles together to reveal the edge configuration of the solution attractor. What is the proper number for K? In our experiments, we found that $K = 6n$ is the magic number. The union of the edge configurations of at most $6n$ random initial tours can generate the edge configuration of the entire solution space (i.e., all cells of the matrix E can be hit by these initial tours). The core structure (the set of the globally superior edges and the G-edges) of the edge configuration of the constructed solution attractor becomes unique and fixed when the number of search trajectories $K \geq 6n$.

Another related question is "how many moves a local search trajectory has to make before it reaches a real locally optimal tour?" So far we do not have an answer to this question. We even do not know any nontrivial upper bounds on the number of moves that may be needed to reach local optimality [14–17]. In practice, we are rarely able to find a true locally optimal point because we simply do not allow the local search process run enough long time. We usually let a search trajectory run a predefined number of iterations, accept whatever solution it generates, and treat it as a locally optimal solution. Therefore, the size of the constructed solution attractor depends not only on the problem structure and the neighborhood function used in the local search process but also on the amount of search time invested in the local search process. If we spend more time in the local search process ($t_2 > t_1$), the resulting constructed solution attractor should be smaller ($A_2 < A_1$), as illustrated in **Figure 6**.

Let M_S be the edge configuration of the solution space S, M_A the edge configuration of the true solution attractor A, and f^t the t-iterate of the search function f in the local search process on K search trajectories, and then it follows easily that M_A is equal to the intersection of the nested sequence of forward edge sets:

$$M_S \supset f(M_S) ... \supset f^t(M_S) ... \supset M_A \qquad (6)$$

Therefore, at any search time t before the K search trajectories reach their true end points, the edge configuration of the true solution attractor M_A is always a subset of the edge configuration of the constructed solution attractor $f^t(M_S)$, and thus the constructed solution attractor is always larger than the true solution attractor.

What is the relationship between the size of the constructed solution attractor and the size of the given problem? So far there is no theoretical or analytical tool available in the literature that can be used to answer this question. We have to depend on empirical results to lend some insights. If the size of the constructed attractor increases exponentially with the size of the problem increases, the ABSS still does not fundamentally reduce the computational complexity of the problem.

Figure 6.
The size of a constructed solution attractor is also determined by the time spent in the local search process.

The ABSS consists of two search phases: the local search phase that construct the solution attractor (from line 5 to line 10 in the ABSS algorithm) and the exhausted search phase that find the best tour in the solution attractor (line 11). For the TSP, the solution space can be represented by a search tree. The local search phase actually performs the task of pruning off the edges that cannot possibly be included in the globally optimal tours. When the first edge is discarded by all K search trajectories, $(n - 2)!$ tours that go through this edge are excluded in the search space of the exhausted search phase. Each time an edge is removed, the search space of the exhausted search phase is reduced by a factor. In such a way, the number of combinatorial branching possibilities for the exhausted search can be exponentially reduced. Decades of research and empirical evidence have found that heuristic local search algorithms converge very quickly, within low-order polynomial time [14]. When majority of the edges are removed, a huge number of possible tours in the solution space are removed from consideration in the exhausted search phase. In this way, the computational complexity of the problem is significantly reduced. In our experiments, the local search process can remove over 70% of edges in the matrix E in a number of iterations bounded by a linear polynomial time. Therefore, the local search phase in the ABSS can be done in $O(n^2)$. **Figure 7** shows the result of one of our experiments. All other similar experiments reveal the same pattern. All our experiments used the 2-opt local search technique because the 2-opt has the smallest expected number of local optima [14]. The experiments were carried out on a PC with 2.60 GHz Intel® Core(TM)i7-3687U CPU, running under Microsoft Windows 7 Enterprise. The ABSS algorithm was coded in Microsoft Visual Basic 2012. In this experiment, we generated 10 unimodal TSP instances in the size from 1000 to 10,000 nodes with 1000-node increment. For each instance, the search system generated $k = 6n$ search trajectories. First, we let each search trajectory stop when no improvement was made during 10,000 iterations, no matter the size of the problem (viz., fixed search time). We counted the number of tours in the constructed solution attractor for each instance. Next we ran the search system again on these instances. This time we made each search trajectory stop when no improvement was made during $10n$ iterations (varied search time 1) and $100n$ iterations (varied search time 2), respectively. Then we counted the number of

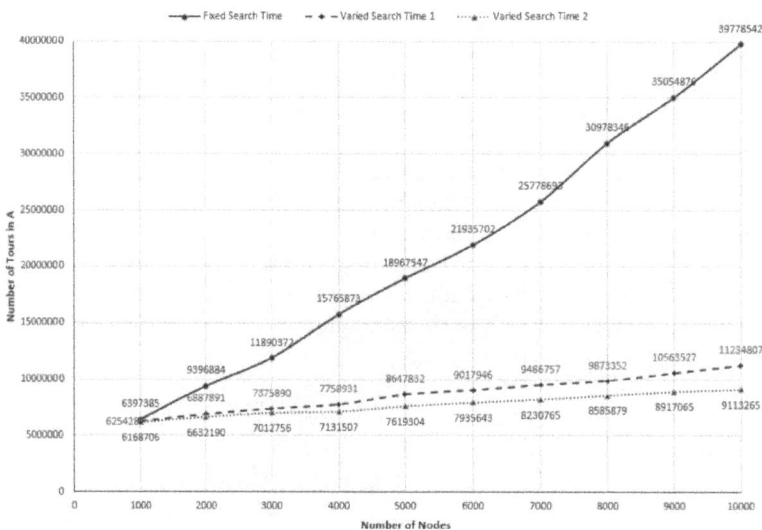

Figure 7.
The relationship between the size of the constructed solution attractor and the size of the problem.

tours in the constructed solution attractor for each instance. As illustrated in the chart of **Figure 7**, all curves appear to be linear, and the varied-search-time curves have much flatter slope because longer local search time leads a smaller solution attractor.

After the local search phase, majority of unnecessary branches have been cut off from the search tree. Usually, when using tree search enumerative algorithm, the effective branching factor is used to measure the computing complexity of the algorithm. An *effective branching factor b** is the number of successors generated by a typical node for a given search tree problem. We use the following definition to calculate effective branding factor b^* in the exhausted search phase:

$$N = b^* + (b^*)^2 + ... + (b^*)^n \qquad (7)$$

where N is total number of nodes generated from the origin node and n is the size of the TSP instance, representing the depth of the tree. We conducted several experiments on different TSP instances. The tree search process always starts from node 1 (the first row of the matrix E). N is the total number of nodes that are processed to construct all valid and invalid tours in the matrix E from the node 1. N does not count the node 1 (the origin node), but includes node 1 as the end node of a valid tour. **Figure 8** shows the result of one experiment, using the same instances and setting reported in **Figure 7**. The effective branching factors in all our experiments are very small, all less than 2. This result indicates that the edge configuration of the solution attractor presents a tree with extremely sparse branches, and the degree of sparseness does not change as the problem size increases if we properly increase local search time for a larger instance. It also indicates that the exhausted search phase is polynomial time if we polynomially increase local search time for larger instances. Therefore, the tree represented by the edge configuration of the constructed solution attractor has a manageable size that can be searched completely in $O(n^2)$.

The ABSS is a global optimization system. The goal of a global optimization system is to find all absolute best solutions in the solution space. There are two major tasks in a global optimization system: (1) finding all globally optimal points in the solution space and (2) making sure that they are globally optimal. To complete these tasks, the global optimization system should meet the following requirements: (1) its search behavior should be globally convergent, (2) it should be deterministic

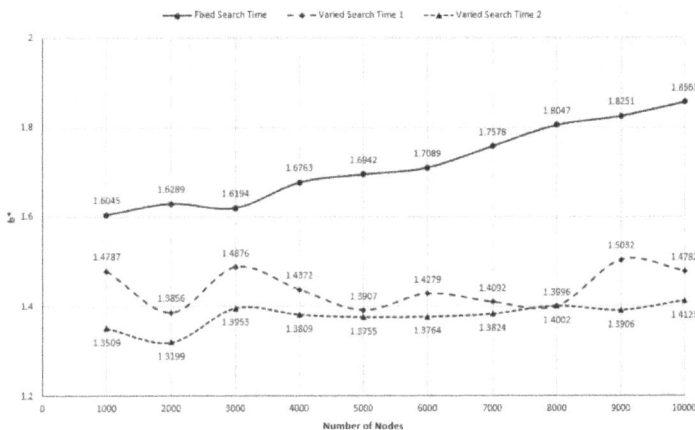

Figure 8.
The b* *values for different problem size* n.

and has a rigorous guarantee for finding all globally optimal solutions without excessive computing burden, and (3) it should have a self-evident optimality criterion.

In the ABSS, two different search phases have different search objectives. The objective of the local search phase is "searching for most promising tours in the solution space." It tries to provide an answer to the question "In which small region of the solution space is the best tour located?" The objective pursued by the exhausted search phase is "finding the best tour among the most promising tours." It tries to provide an answer to the question "In this small region, which tour is the best one?" Putting these two objectives together, the ABSS tries to provide an answer to the question "Which tour is the best tour in the solution space?"

The ABSS combines beautifully two crucial aspects in search: exploration and exploitation. In the local search phase, K search trajectories explore the solution space independently and individually to collect the edges for constructing the solution attractor. The K search trajectories create and maintain diversity from beginning to the end. Randomization in the local search process makes the local search process become a randomized process. A search trajectory changes its edge configuration according to the objective function and its neighborhood structure. The local search phase actually uses the Monte Carlo simulation to sample locally optimal tours. Monte Carlo simulation is defined as simulations used to model the probability of different outcomes in a process that cannot easily be predicted due to intervention of random variables. The essential idea of Monte Carlo method is to use randomness to solve problems that might be deterministic in principle. In the ABSS, K search trajectories start a sample of initial tours from uniform distribution over the solution space and, through a randomized local search process, generate a sample of locally optimal tours that are uniformly distributed in the constructed solution attractor. Therefore the edge configuration of the solution attractor is constructed through this Monte Carlo sampling process. The distribution of the hit edges in the matrix E converges to a small set of edges, and the set of the edges is statistically fixed. This fixed edge configuration is not sensitive to the selection of K search trajectories. Convergence and stability are two desirable properties of the solution attractor: all search trajectories will converge to the solution attractor and remain there forever. The ability of K search trajectories to explore the entire solution space and thus collect all globally superior edges and G-edges can help the ABSS achieve its required function—finding all globally optimal tours.

The global convergence and deterministic property of the search trajectories make the ABSS converge in solution, that is, the ABSS always find the same set of the best tours. This argument was empirically confirmed in our experiments. For a given TSP instance, we repeated the same search process on the same instance many times, each time using a different set of K search trajectories, and the search system always generates the same set of the best tours in all trials. **Table 1** shows the result of one experiment. This experiment generated two TSP instances Q_1 and Q_2 with $n_1 = 1000$ and $n_2 = 10000$ nodes. The ABSS ran each instance 15 times, each time using a different set of $K = 6n$ search trajectories. The ABSS found the same single best tour in all 15 trials for Q_1 and the same set of three best tours in all 15 trials for Q_2. The three best tours for Q_2 have the same cost value but with different edge configurations. It is clear that Q_1 is a unimodal TSP instance and Q_2 is a multimodal instance having three optimal tours in its solution space. If any trial had generated a different set of the best tours, we could immediately make a conclusion that the best tours in the constructed solution attractor may not be the globally optimal tours. From the experimental and practical perspective, the fact that the same set of the best tours was detected in all trials provides a significant empirical evidence of the optimality of these tours.

Trial #	Number of tours in A	Range of tour cost	Number of best tours in A
1000 nodes (Q_1)	(6000 initial tours)		
1	5,703,833	[3926, 4437]	1
2	5,703,785	[3926, 4521]	1
3	5,703,479	[3926, 4509]	1
4	5,703,829	[3926, 4495]	1
5	5,703,868	[3926, 4540]	1
6	5,703,499	[3926, 4500]	1
7	5,703,253	[3926, 4556]	1
8	5,703,791	[3926, 4488]	1
9	5,703,742	[3926, 4498]	1
10	5,703,990	[3926, 4551]	1
11	5,703,637	[3926, 4526]	1
12	5,703,457	[3926, 4536]	1
13	5,703,642	[3926, 4534]	1
14	5,703,626	[3926, 4546]	1
15	5,703,727	[3926, 4522]	1
10,000 nodes (Q_2)	(60,000 initial tours)		
1	9,428,645	[81,967, 85,287]	3
2	9,428,571	[81,967, 84,979]	3
3	9,428,032	[81,967, 85,286]	3
4	9,429,004	[81,967, 85,365]	3
5	9,428,625	[81,967, 85,348]	3
6	9,428,819	[81,967, 85,345]	3
7	9,428,815	[81,967, 85,232]	3
8	9,429,021	[81,967, 85,254]	3
9	9,428,950	[81,967, 85,320]	3
10	9,428,847	[81,967, 85,286]	3
11	9,428,749	[81,967, 85,036]	3
12	9,428,978	[81,967, 85,248]	3
13	9,428,767	[81,967, 85,076]	3
14	9,428,933	[81,967, 85,223]	3
15	9,428,799	[81,967, 85,337]	3

Table 1.
Tours in solution attractor for 1000-node and 10,000-node TSP instances.

One factor that makes the TSP difficult to solve is that we have not found a simple optimality criterion to decide whether or not a locally optimal tour is also a globally optimal tour. Selecting the best tour among a set of tours and knowing it is the best one are the full challenges of the TSP. A brute-force algorithm that sorts through all tours in the solution space can be certain that it meets the challenge. However, it lacks practical efficiency. For a TSP instance, there are an unknown number of globally and locally optimal tours. The ABSS uses a simple and practical optimality criterion: the best tours in the set of all locally optimal tours are the globally optimal tour. In fact, this criterion is the necessary and sufficient condition for a locally optimal tour to be a globally optimal tour. In the ABSS, the local search phase identifies the solution attractor, and no tour outside the solution attractor can be better than any tour inside. Then the exhausted search phase examines all tours in the solution attractor and finds the best tours. In fact, this optimality criterion describes how the ABSS models and solves the TSP.

For a tour $s_i \in S$, its neighborhood $N(s_i) \subset S$ is defined, consisting of all tours that can be reached from s_i in one single transition. A locally optimal tour s' satisfies $f(s') \leq f(s)$ for all $s \in S \cap N(s')$. A solution attractor A consists of all locally optimal tours. A best tour s^* in a solution attractor satisfies $f(s^*) < f(s')$ for all $s' \in A$. The best tour $s^* \in A$ satisfies the following conditions, which allow the propagation of the minimum properties of s^* in the solution attractor A to the whole solution space S, that is, $f(s^* \in A) < f(s)$ for all $s \in S$:

1. $f(s') \leq f(s)$ for all $s \in S \cap N(s')$

2. $A \ni s'$ for all $s' \in S$

3. $f(s^* \in A) < f(s')$ for all $s' \in A$

4. $\min\limits_{s \in S} f(s) = \min\limits_{s \in A} f(s)$

5. $\lim\limits_{t \to \infty} f(s^*) = s^*$

6. Conclusions

For the TSP, the computational complexity is associated with the combinatorial explosion of potential solutions in the solution space. If we accept the argument that the number of tours in the solution space indicates the difficulty of the TSP, then the fact that the solution space can be significantly reduced to a small solution attractor means that the difficulty of the TSP can be dramatically reduced. The novel perspective of solution attractor in a local search system for the TSP gives us an opportunity to overcome combinatorial complexity. The solution attractor shows us where the best tour can be found in the solution space. If we concentrate the exhausted search effort in this much smaller region, the number of possibilities in search space is no longer prohibitive. Our experiments showed that the ABSS can significantly reduce the computational complexity for the TSP and thus can solve the TSP much efficiently with global optimality guarantee. The ABSS is an obvious finite algorithm in computing complexity of $O(n^2)$ and space requirement of $O(n^2)$ for the TSP. This suggests that the TSP might not be as complex as we might have expected.

The edge matrix E is the data structure that is defined by the TSP naturally and is used in the ABSS to separate the solution attractor from the entire solution space. In the ABSS, the combination of an efficient local search process, a powerful data structure (the matrix E), and an exhausted search process provides a highly effective and efficient search system. If some other NP-hard problems have the same nice data structure that can be used to reduce the search space, these problems can also be solved in polynomial time.

This chapter focuses on the solution attractor of the local search system for the TSP. Does it appear to be technical archetypes for other combinatorial optimization problems? Each optimization problem has its own specifics and data structure. In order to fully understand the search process for a particular problem, we must put our attention to the data structure that is defined by the problem. The combination of a proper data structure and simple search strategy can make the highly complex solution space become tractable and lead to more knowledge about the problem and provide opportunities for new algorithmic designs.

The TSP is the most prominent problem in NP-hard problems. It is hoped that this chapter will serve as a pioneer in this field and bring more and better works from other researchers and practitioners. The ultimate goal of this chapter is to encourage readers to take up their own pursuit of interesting problem-by-problem methods for attacking diverse optimization problems.

The solution attractor theory provides some important insights into the power of efficient computations and a line of reasoning that may lead to a proof in the near future about P vs. NP problem. The P vs. NP problem is an important computational issue in nearly every scientific discipline [18]. It is about how efficient we can search through a huge number of possibilities. Computational complexity theory suggests that there are limits of the power of general-purpose optimization techniques. Majority of people are in favor of $P \neq NP$ because we totally lack fundamental progress in the area of enumerative search [19]. What are these limits? If we design a search algorithm that fully utilizes the natural structure of the problem, like the edge matrix E of the TSP, we may be able to remove some constraint on our road.

Author details

Weiqi Li
University of Michigan – Flint, Flint, USA

*Address all correspondence to: weli@umich.edu

IntechOpen

References

[1] Ausiello G, Crescenzi P, Kann V, Marchetti-sp G, Spaccamela M. Complexity and Approximation: Combinatorial Optimization Problems and their Approximability Properties. New York: Springer; 2003

[2] Rego C, Gamboa D, Glover F, Osterman C. Traveling salesman problem heuristics: Leading methods, implementations and latest advances. European Journal of Operational Research. 2011;**211**:427-441

[3] Korte B, Vygen J. Combinatorial Optimization: Theory and Algorithms. New York: Springer; 2012

[4] Alligood KT, Sauer TD, Yorke JA. Chaos: Introduction to Dynamical System. New York: Springer; 2000

[5] Brin M, Stuck G. Introduction to Dynamical Systems. Cambridge: Cambridge University Press; 2016

[6] Brown RA. Modern Introduction to Dynamical System. New York: Oxford University Press; 2018

[7] Dénes A, Makay G. Attractors and basins of dynamical systems. Electronic Journal of Qualitative Theory of Differential Equations. 2011;**20**:1-11

[8] Milnor J. On the concept of attractor. Communications in Mathematical Physics. 1985;**99**:177-195

[9] Milnor J. Collected Papers of John Milnor VI: Dynamical Systems (1953–2000). Washington, DC: American Mathematical Society; 2010

[10] Li W. Dynamics of local search trajectory in traveling salesman problem. Journal of Heuristics. 2005;**11**: 507-524

[11] Li W, Feng M. The solution attractor of local search in traveling salesman

problem: Concept, construction and application. International Journal of Metaheuristics. 2013;2:201-233

[12] Li W, Li X. The solution attractor of local search in traveling salesman problem (part 2): Computational study. International Journal of Metaheuristics. 2019;7:93-126

[13] Shannon CE. A mathematical theory of communication. Bell System Technical Journal. 1948;27:623-656

[14] Aarts E, Lenstra JK. Local Search in Combinatorial Optimization. Princeton: Princeton University Press; 2003

[15] Chandra B, Karloff H, Tovey C. New results on the old k-opt algorithm for the traveling salesman problem. SIAM Journal on Computing. 1999;28: 1998-2029

[16] Fischer ST. A note on the complexity of local search problems. Information Processing Letters. 1995;**53**: 69-75

[17] Grover LK. Local search and the local structure of NP-complete problems. Operations Research Letters. 1992;**12**:235-243

[18] Fortnow L. The Golden Ticket – P, NP, and the Search for the Impossible. Princeton: Princeton University Press; 2013

[19] Fortnow L. The status of the P versus NP problem. Communications of the ACM. 2009;**52**:78-86

Chapter 4

Accelerating DNA Computing via PLP-qPCR Answer Read out to Solve Traveling Salesman Problems

Fusheng Xiong, Michael Kuby and Wayne D. Frasch

Abstract

An asymmetric, fully-connected 8-city traveling salesman problem (TSP) was solved by DNA computing using the ordered node pair abundance (ONPA) approach through the use of pair ligation probe quantitative real time polymerase chain reaction (PLP-qPCR). The validity of using ONPA to derive the optimal answer was confirmed by *in silico* computing using a reverse-engineering method to reconstruct the complete tours in the feasible answer set from the measured ONPA. The high specificity of the sequence-tagged hybridization, and ligation that results from the use of PLPs significantly increased the accuracy of answer determination in DNA computing. When combined with the high throughput efficiency of qPCR, the time required to identify the optimal answer to the TSP was reduced from days to 25 min.

Keywords: DNA computing, traveling salesman problem, ordered node pair abundance, pair ligation probe-qPCR, PLP-qPCR

1. Introduction

The traveling salesman problem (TSP) computes the shortest route on the arcs of a network that visits a given set of nodes (cities) before returning to the starting point [1–6]. In many cases, *in silico* computers are incapable of quickly determining an exact solution of these nondeterministic polynomial (NP) problems because a linear increase in the number of variables leads to a factorial increase in the number of potential solutions. Although advanced heuristic methods have increased the ability of *in silico* computers to provide approximate answers to NP problems [7], they lack the massive parallelism and data storage required to find exact solutions. DNA computing, which uses the hybridization of DNA molecules as a means to make computations [8–15], is particularly well-suited to solve computationally intense NP problems such as the TSP because multiple sequences of DNA in a solution can hybridize simultaneously, thereby performing massively parallel computing.

Several technical limitations have prevented DNA computing from reaching its full potential. Although the computation can occur within seconds, current applications of DNA computing are largely limited by time-consuming and labor-intensive answer sorting and determination processes. For example, a gradient PCR procedure that involved a series of PCR reactions with a variety of primer pair combinations was used to determine the solution to a 7-city directed Hamiltonian circuit problem [8]. Other answer determination methods include DNA sequencing [16], and denaturation temperature gradient-polymerase chain reaction (DTG-PCR) associated with denaturing gradient gel electrophoresis (DGGE) and/or temperature gradient gel electrophoresis (TGGE) [10, 17, 18]. Neither DNA sequencing nor DGDC/TGDC can be used to identify the optimal answer to a TSP because the feasible answers represent both optimal and suboptimal answers that differ only in the order in which the components were ligated when the answers were formed, and are flanked by the sequences that encode the start and end nodes. The dependence of the discriminatory powers of DTG-PCR, DGGE, and TGGE on the diversity of oligonucleotide components, especially the G and C content [19, 20] also limits their applicability for answer determination of NP-complete problems.

Xiong et al. [12] developed the ordered node pair abundance (ONPA) approach to identify the optimal answer of an asymmetric, fully-connected 15-city TSP, the largest problem solved to date, using DNA computing. In that study, 20-mer DNA sequences that specifically encoded the nodes and arcs in the network were added to the reaction mixture in the presence of ligase. The sequence of each arc was capable of linking two node sequences in a specific order via hybridization of the last 10 bases of the prior node to the first 10 bases of the subsequent node. The efficiency of each arc in the network was encoded as the concentration of DNA for that arc using saturating concentrations of DNA for each node. Ligation of these sequences then formed covalently linked ordered node pairs (ONPs) that assembled into answer sequences representing tours through the network, such that the optimal answer would be formed in greatest abundance.

The optimal answer to the 15-city TSP was identified by ONPA after infeasible answer sequences were removed by electrophoresis and magnetic bead separation [11, 12]. For each ONP examined by ONPA, probes were designed to hybridize to the 3′-end of the prior node and the 5′-end of the subsequent node in the ordered pair as well as a complementary pair for use in ligation chain reaction (LCR). Each pair of probes was then subjected to the same number of LCR cycles in the presence of an aliquot of answer sequences, and the relative abundance of LCR product was quantified by PAGE band intensity. This was repeated for every possible combination of ONPs to identify the ones present in greatest abundance. Although this provided a clear determination of the optimal answer, the accuracy and precision of the computation were limited by the error inherent in determining DNA abundance via bands in a PAGE gel and by the fact that LCR is an end-point assay. In addition, the procedure required days to complete.

Quantitative real-time PCR (qPCR), which uses an increase in fluorescent signal to monitor increases of PCR amplicons, has proven to be a fast, precise and reproducible method to rapidly quantify the relative abundance of many nucleic acid sequences. This results in part because qPCR is not an end-point assay. Instead, the cycle at which a positive signal is first consistently detectable, termed the cycle threshold (C_t), is proportional to the initial content of a given template in the sample. Higher initial target contents give rise to earlier detectable increases in signal, resulting in lower C_t values. Simultaneous amplifications of multiple DNA targets in a single reaction tube can be achieved by multiplex qPCR, which maximizes throughput and decreases the time required to collect information. Ibrahim et al. [21] reported a qPCR approach to readout answers to a Hamiltonian path problem. However, the protocol required multiple combinations of qPCR amplifications and was ultimately dependent upon the use of *in silico* information processing to determine the answer.

Answer determination via a direct qPCR amplification of the entire length of TSP answer sequences is impossible. This is due to the intra-molecular heterogeneity of answers sharing the same starting and the ending nodes, which are the same limitations that eliminate the use of DTG-PCR. The application of qPCR for the quantification of the short 20-mer sequences using ONPA for answer determination of a TSP has also not been possible because standard PCR methodologies require two DNA primers that flank the region of the DNA sequence to be amplified. However, we developed a pair ligation probe (PLP)-dependent qPCR system (PLP-qPCR) that is capable of rapidly quantifying the abundance of short DNA sequences under the multiplex conditions [22]. Upon hybridization to its 20-mer target DNA, the PLP becomes circularized by ligation. This allows the abundance of adjacent short sequences in a DNA strand to be transformed into the copy number of circularized PLP that can be amplified by qPCR.

We have now solved an asymmetric, fully-connected 8-city TSP by the ONPA approach using PLP-qPCR without the need for *in silico* information processing. The validity of using ONPA to derive the optimal answer was confirmed using a reverse-engineering method to reconstruct the complete tours in the feasible answer set from the measured ONPA. The high specificity of the sequence-tagged hybridization and ligation that results from the use of PLPs significantly increased the accuracy of answer determination in DNA computing. When combined with the high throughput efficiency of qPCR, the time required to identify the optimal answer to the TSP was reduced from days to 25 min.

2. Materials and methods

2.1 Oligonucleotide design and construction

Oligonucleotide sequences were designed using Primer Express Version 2 for Windows (Applied Biosystems) and were synthesized by Invitrogen Inc. Nodes labeled B through H were represented by synthetic 20-mer sequences of DNA except for the starting and ending nodes (A_{start} and A_{end}) that were comprised of 30-mers (**Table 1**). Node sequences were chosen to minimize cross hybridization, and nodes B through H had a GC content from 30% to 35% such that melting points varied from 60.6 to 62.0 °C, while the A_{start} and A_{end} sequences contained GC contents of 66% and 72%, respectively. None of the arc sequences were complementary to the first 15-mer sequence for the A_{start} and the last 15-mer sequence for the A_{end}. Thus, incorporation of A_{start} or A_{end} into an answer sequence prevented further extension of that end of the answer sequence. The start and end sequences also served as primer sequences for downstream amplification by PCR, which provided the capability to increase the amount of answer sequences. The longer A_{start} and A_{end} sequences with higher GC content were important for downstream PCR amplification with enhanced fidelity and improved efficiency for answer purification. Any two node sequences could be linked together by a 20-mer arc sequence composed of two 10-mers that were complementary to the last and first 10 nucleotides in the sequences of the former and latter nodes, respectively (**Figure 1**). Arcs were made to complement every combination of nodes with the exception of the respective 5' and 3' ends of A_{start} and A_{end}. Oligo sequences (**Table 2**) designed for the PLP-mediated qPCR assay followed the protocol of Xiong and Frasch [23]. All oligonucleotides were purified by PAGE under denaturing conditions [12].

Node	Length	Sequence[a]	GC (%)	T_M (°C)
A_S[e]	30	TCTGCGGGCGGACAGACATGGTTAGCGGCC[b,c]	66	70.0
B	20	TTTACGTCTACCATATCTAT[d]	30	61.2
C	20	ATAGCAACACTACATATGTC	35	61.6
D	20	TCGACTAATTCGTACTTATA	30	61.6
E	20	CCTGATACAAGTACTAAGTA	35	61.6
F	20	GCGTAAGGATATTTATACAA	30	61.6
G	20	GTTTGTTTAGTCCATCATTA	30	61.7
H	20	AGCATTATTTCTTCCAAATA	25	61.7
A_E[e]	30	CTACTGCCGCCGCCGGGTAGACGGCTCGGA	72	72.0

[a] All sequences read in the 5′ to 3′ direction.
[b] Blue sequences can hybridize to the 3′-ends of arc sequences.
[c] Green sequences are unable to hybridize to arc sequences but serve as primer sequences for PCR.
[d] Black sequences can hybridize to the 5′-ends of arc sequences.
[e] A_S is the sequence for node A_{start}; the A_E is the sequence for the node A_{end}.

Table 1.
Node sequences used in the calculation.

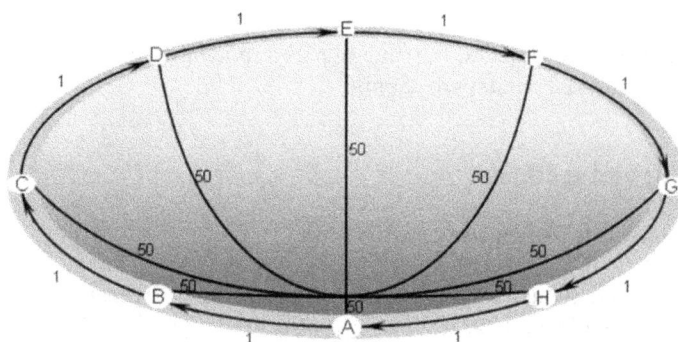

Figure 1.
Graphical representation of the 8-city TSP solved by DNA computing. The optimal tour through the network visits the nodes in alphabetical order. All other tours include arcs that are 100-fold less efficient, and cross at a common point.

2.2 Answer formation and purification

Hybridization was performed by heating the answer formation reaction medium (AFRM) to 92°C for 4 min, followed by a programmed annealing process at a cooling rate of 1°C per min to 8°C. The AFRM was composed of all relevant oligonucleotides in T4 Ligase Buffer (Fermentas). Ligation was performed by incubating at 8°C for ~16 h after addition of 5 Weiss U of T4 ligase, 20 mM DTT, and 10 mM ATP to the AFRM.

Ligation products containing all answer sequences were separated by 6% denaturing PAGE containing 8 M urea and 15% formamide at 95 V for about 3 h. The gels were visualized with UV light after staining with ethidium bromide (1 mg/ml) for 10 min. Fragment sizes were determined by comparison with mobility of a 20-bp DNA ladder (Bayou BioLabs). The 190 bp band was excised from the gel that contained sequences consistent in length to those of feasible answers (i.e. containing one copy of each node sequence starting and ending with A_{start} and A_{end}). These sequences were amplified by PCR using PCR primers for A_{start} and

DNA sequence	
qPCR reporter-identifying sequences	
NED-TAqMan® spacer	CGCGGATGTCGGTCAGCCGAGTCTACCCAGCGCGCCACTATCGCCATCAGGCAGC
TaqMan-MGB® reporter	
NED	NED-TCGCCATCAGGCAGC-MGBNFQ
Pair ligation arms for node recognition	
5′-B′	ATAGATATGG
5′-D′	TATAAGTACG
5′-F′	TTGTATAAAT
3′-C″	GTGTTGCTAT
3′-E″	TTGTATCAGG
3′-G″	CTAAACAAAC
qPCR primers	
P_r	TAGACTCGGCTTGACCGACATCCGCG
P_f	TCTACCCAGCGCGCCAC

Table 2.
Oligonucleotide sequences for the 5′ and 3′ arms of the pair ligation probes, the forward (P_f) and reverse (P_r) primers, and the TaqMan reporters used for PLP-qPCR.

A_{end}, and subject to sequential magnetic affinity purification steps as reported by Spetzler et al. [11, 12] to purify feasible answers.

2.3 Preparation of target-specific pair ligation probes

Each PLP consisted of one 55-mer core and two 10-mer target-specific sequences that comprised the 5′ and 3′ arms located at ends of the core (**Table 2**). The core sequence contained the forward (P_f 19-mer) and reverse (P_r 21-mer) PCR primer-binding sequences, and a qPCR reporter-identifying sequence known as a TaqMan spacer for use with the TaqMan-MGB® (Applied Biosystems) NED reporter dye (λ_{max} = 580 nm). The PLPs specific for each ONP in the answer sequences were made from core and arm components by ligation as per Xiong and Frasch [23].

2.4 Circularization of the PLP and the qPCR assay

Aliquots of the purified answer sequences containing ~2 pmol DNA were denatured and annealed with 20 pM of linear PLP. The hybridized PLPs were circularized by ligation at 10°C overnight. The ligation reaction mixture contained 2 μl of 10× ligation buffer (Fermentas), 50 mM DTT, and 5 Weiss U of T_4 DNA ligase, in a final volume of 20 μl. After ligation, 2 μl of ligation product was added to 18 μl of the exonuclease mixture that contained 10 mM Tris-HCl, pH 9.0, 5 mM $MgCl_2$, 0.1 mg per ml of BSA, 10 U Exonuclease I and 10 U Exonuclease III to remove any remaining linear PLPs. The samples were incubated at 37°C for 2 h followed by inactivation at 65°C for 20 min.

Quantitative real-time PCR (qPCR) assays were performed in a 96-well, closed plate using the AB 7500 Fast RT-PCR System (Applied Biosystems). In a typical qPCR assay, 20 μl of qPCR reaction mixture that contained 10 μl of 2× TaqMan® Fast universal PCR Master Mix including the ROX fluorescent reporter as a passive reference, 900 nM of each of the P_f and P_r primers, 0.25 μM of the NED

TaqMan-MGB® reporter, and ~2 ng of the circularized PLP were used to determine the abundance of each ordered node pair. Thermal cycling profiles for qPCR included heating at 95°C for 20s followed by 45 cycles with 95°C for 3s and 60°C for 30s.

During the PLP-qPCR assay, DNA amplification was monitored by quantitatively analyzing fluorescence emissions using an ABI prism sequence detector. The intensity of the NED reporter dye was measured against the ROX internal reference dye signal to normalize for non-PCR-related fluorescence fluctuations that occurred from well to well. The threshold cycle (C_t) represented the refraction cycle number at which a positive amplification reaction was measured and was set at 10 times the standard deviation of the mean baseline emission calculated from PCR cycle 5–15.

2.5 Linear programming (LP) model of tour frequency from ONPA data

The model is based on LP approaches to data fitting [24–27], and solves for the tour frequencies X_t that best fit the observed ONPA F_{ij}.

$$\text{Minimize} \sum_{i=1}^{8} \sum_{j=1}^{8} \left(d_{ij}^+ + d_{ij}^- \right) \tag{1}$$

$$\text{Subject to:} \sum_{t=1}^{5040} h_{tij} X_t + d_{ij}^+ - d_{ij}^- = F_{ij} \quad \text{for all } i,j \tag{2}$$

$$X_t, d_{ij}^+, d_{ij}^- \geq 0 \quad \text{for all } t,i,j \tag{3}$$

where d_{ij}^+ = positive deviation from observed frequency of arc ij; d_{ij}^- = negative deviation from observed frequency of arc ij; X_t = estimated frequency of tour t; h_{tij} = 1, if tour t contains ONP ij; 0 otherwise; F_{ij} = observed frequency of arc ij.

The number of possible tours t that begin and end with node A, and visit each node only once is P_7^7 = 7! = 5,040. Since the number of DNA molecules of tour t that were made by the DNA computer were not explicitly measured, one X_t variable is generated for each tour t, representing how many DNA molecules of tour t were likely to have been made, given the ONPA. The goal was to solve for the best-fitting values of X_t. Constraints **2** are written once for each ONP ij. The first expression on the left-hand side of the constraint, $\sum_{t=1}^{5040} h_{tij} X_t$, sums all occurrences of arc ij within all tours t. Constraints **2** then add a positive deviation or error (d_{ij}^+) and subtract a negative deviation (d_{ij}^-) to the left-hand side and equate it to the observed ONPA F_{ij}. Thus, if there is no set of X_t values that can perfectly reproduce the observed ONPA, the deviation variables are forced to take up any slack or surplus. The objective function **1** minimizes the sum of the deviations so as to find the best-fitting tour frequencies X_t for the observed ONPA F_{ij}. The deviations are separated into positive and negative components because linear programming cannot handle absolute values explicitly. The deviation variables, like the tour frequency variables, are constrained by **3** to be non-negative. If the deviation variables were permitted to be either positive or negative, minimizing their sum would not provide a good fit because the objective would reward large negative deviations pushed toward $-\infty$.

3. Results

The computation to find the optimal solution to an asymmetric, fully-connected 8-city TSP was defined by the distance matrix in **Table 3**. The problem was

Prior	Subsequent node								
Node	A_S	B	C	D	E	F	G	H	A_E
A_S	*a (0)b	1c (100)b	100 (1)	100 (1)	100 (1)	100 (1)	100 (1)	100 (1)	100 (1)
B	* (0)	* (0)	1 (100)	100 (1)	100 (1)	100 (1)	100 (1)	100 (1)	100 (1)
C	* (0)	100 (1)	* (0)	1 (100)	100 (1)	100 (1)	100 (1)	100 (1)	100 (1)
D	* (0)	100 (1)	100 (1)	* (0)	1 (100)	100 (1)	100 (1)	100 (1)	100 (1)
E	* (0)	100 (1)	100 (1)	100 (1)	* (0)	1 (100)	100 (1)	100 (1)	100 (1)
F	* (0)	100 (1)	100 (1)	100 (1)	100 (1)	* (0)	1 (100)	100 (1)	100 (1)
G	* (0)	100 (1)	100 (1)	100 (1)	100 (1)	100 (1)	· (0)	1 (100)	100 (1)
H	* (0)	100 (1)	100 (1)	100 (1)	100 (1)	100 (1)	100 (1)	* (0)	1 (100)
A_E	* (0)	* (0)	* (0)	* (0)	* (0)	* (0)	* (0)	* (0)	* (0)

[a]Arc does not exist.
[b]pmoles of arc molecules input into a final volume of 56 µl are shown in parentheses.
[c]Distances.

Table 3.
Arc distance and corresponding concentration matrices used to make the computation.

designed such that the alphabetical order of the nodes was the optimal tour
(**Figures 1** and **2d**). To accomplish this, the distances were translated into concentrations (1:1 pmol) of the arc sequences that connect the nodes in inverse proportion to the distances (**Table 3**). These sequences were then hybridized with an excess of all node sequences and ligated to form answer strands as summarized in **Figure 2**. Node and arc sequences were asymmetric in design so that two node sequences could become ligated in an order-specific manner to form an ordered node pair (ONP). For example, to compute the formation of the BC-ONP, the last (3′) and first (5′) halves of the sequences for nodes B and C hybridize to the first (3′) and last (5′) halves of the BC arc oligonucleotide, respectively, which then permits ligase to link them covalently.

Given that the concentration of arcs along the upper diagonal of the efficiency matrix was 100-fold that of the other arcs, strands containing node sequences in alphabetical order were anticipated to be produced in highest abundance. A comparable result was obtained previously using a similar approach to solve an asymmetric, fully-connected 15-city TSP [10]. Consequently, this computational method could be reliably used to examine the efficacy of PLP-qPCR in the answer determination step of a DNA computation.

A tour of the nodes was feasible, whether it was optimal or suboptimal, only if it contained one copy of each node sequence flanked by the A_{start} and A_{end} sequences (**Figure 2d–f**). **Figure 3** shows a stepwise summary of the procedure used to make the calculation. The PAGE profile of answer sequences at various stages of purification is shown in **Figure 4**. Selective amplification of answer sequences flanked by A_{start} and A_{end} (Lane 1) was achieved by PCR using primers specific for those sequences (**Figure 2a–f**). This PCR product (Lane 2) included the 190-mer band that was the size required for feasible answers (**Figure 2c–f**), which was then subjected to sequential magnetic bead affinity purification for every node sequence. This insured that all sequences used for answer analysis contained one and only one copy of each node sequence, and thus represented Hamiltonian circuits through the network. The purified sample composed only of answer DNA sequences for feasible tours (i.e. **Figure 2d–f**) appeared as a single 190-mer band (Lane 3) that was used to determine the optimal answer to the computation.

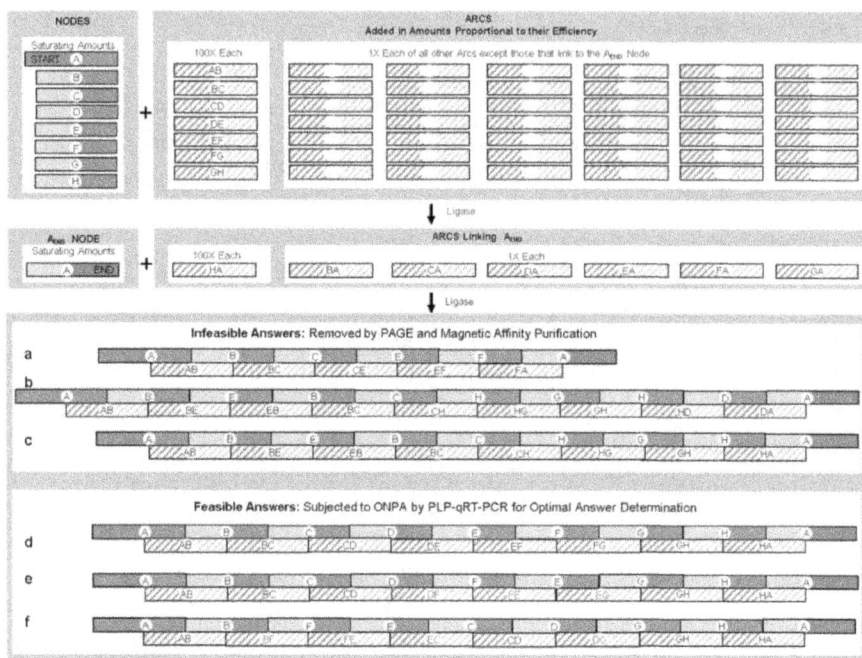

Figure 2.
*Protocol used to compute the 8-city TSP. Answers were formed by the addition of saturating amounts of all nodes and limiting amounts of arcs in inverse proportion to their respective distance. Upon hybridization, nodes and arcs were ligated to form answers. The ligation was accomplished in two stages to increase the probability of forming feasible answers that contained a single copy of each node flanked by the start and end sequences. Infeasible answers were removed by PAGE and by magnetic affinity purification as described in Section 2. Examples of infeasible answer sequences formed include those that are: (a) too short because they lack one or more nodes; (b) too long because they contain multiple copies of nodes; and (c) the correct length but lack at least one node. Feasible answers were subjected to PLP-qPCR for ONPA. Examples of feasible sequences include: (d) the optimal answer sequence with the nodes in alphabetical order; (e) a suboptimal sequence that contains a minimum substitution of 3 arcs from the optimal (see **Table 5**, normalized data row 2), and (f) an even less optimal sequence in which 4 arcs have been substituted from those found in the optimal answer (see **Table 5**, normalized data row 3).*

To identify the optimal answer, the feasible answers (**Figure 4**, Lane 3) were analyzed by the ONPA approach. This was accomplished using PLP-qPCR with the 56 PLPs that were specific to each of the possible ONPs. Each PLP was added to an aliquot of the solution containing the answer set to prepare it for PLP-qPCR. Preparation included the steps summarized in **Figure 5** for the BC-PLP that was used to detect the ONP in which the sequence encoding node B immediately precedes that of node C in a tour. For the BC-ONP, the 5′-arm of the PLP complementary to the last 10 bases of sequence B (designated B′), and the 3′-arm (designated C″) that were complementary to the first 10 bases of sequence C were created [23]. The hybridized PLP was circularized by ligase, after which the PLP was denatured from the target strand and exonuclease was added to eliminate any DNA that had not been circularized. Because the amount of circulated PLP was quantitatively related to the initial content of a target, the target content used in the determination was transformed into the copy number of the circularized PLPs. The presence of a qPCR reporter-identification sequence, and PCR primers in the PLP enabled the use of qPCR to measure the copy number.

Figure 6 shows the PLP-qPCR fluorescence amplification plots as a function of cycle number that determines the amount of each possible ONP in the answer sequences of the 8-city TSP. The ONPA was determined in groups according to the

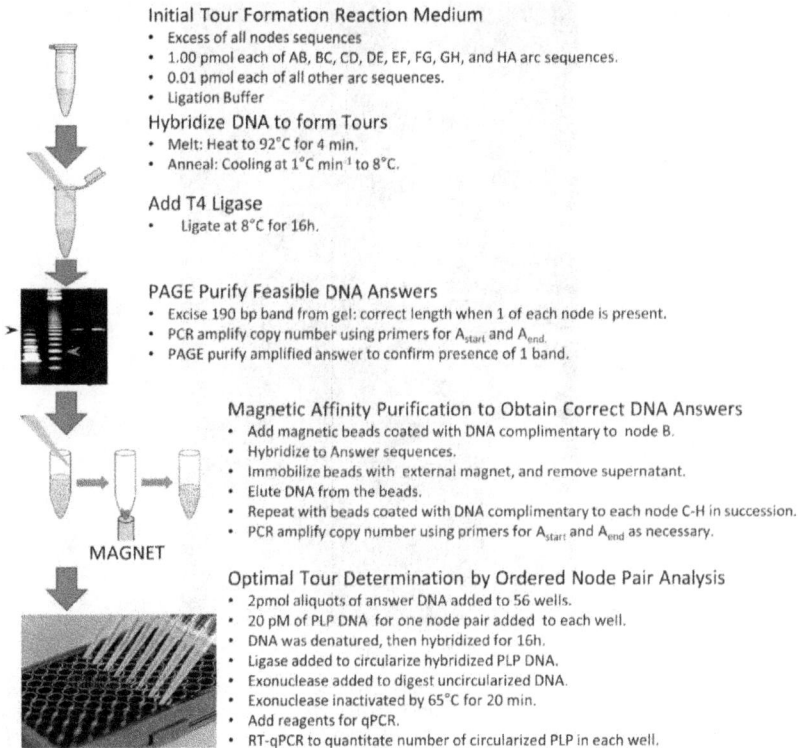

Figure 3.
Graphical representation of the process to make, and purify DNA sequences that represent correct answers, and the use of PLP sequences for ordered node pair analysis to identify the optimal tour. Details of the methods are described in Section 2.

prior node in the ONP such that the relative abundance of any of the ONPs in that group could be compared directly. In each case, it is evident that the first sample in which fluorescence amplification consistently increased above the threshold corresponded to that ONP where the nodes appeared in alphabetical order (i.e. AB, BC, etc.). Thus, it was possible to determine the optimal answer to the 8-city TSP using PLP-qPCR by simple inspection of the raw data. Due to the quantitative nature of qPCR, the C_t values measured from the data in **Figure 3** were used to calculate the copy number of each ONP (**Table 4**). The results are organized so that each row and column define the prior and subsequent nodes of each ONP. Since these measurements were made in groups corresponding to each row, comparisons of the ONPA between rows were achieved by determining the fractional ONPA after normalizing the ONPA present in each row in highest abundance (data in parentheses). The average copy number for suboptimal ONPs was 0.35% of the optimal ONP in each row. However, the suboptimal ONPs were not present in equal abundance. Seven of the suboptimal ONPs were present in amounts that were below the limit of detection. However, the ability to detect these low abundance ONPs was not needed to obtain the optimal answer to the problem solved here. Some of the suboptimal ONPs were formed in much higher abundance than the average. **Figure 7A** shows the number of suboptimal ONPs from **Table 4** as a percentage of the preceding node in the pair. The fraction of suboptimal ONPs was significantly larger in rows C, D, E and F in **Table 4**. This was due to a relatively small number of specific alternate ONPs that included CE (0.2%), DA (0.2%), EC (0.47%), FD (0.5%), and FE (0.47%).

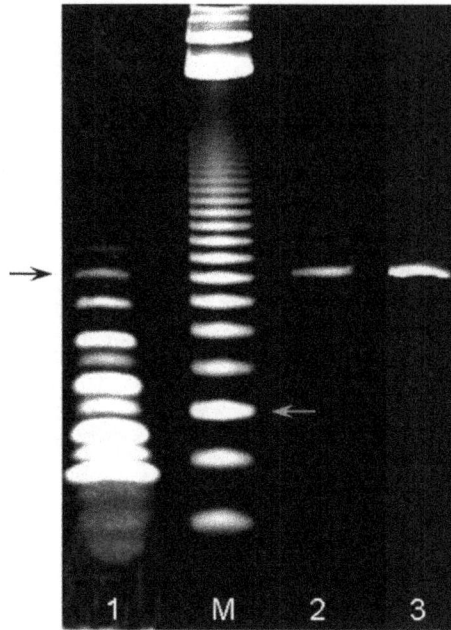

Figure 4.
Profiles of DNA computing products by super denaturing PAGE following amplification by PCR using primers specific to A_S and A_E. 1: Initial product of hybridization/ligation from the answer formation process. Black arrow indicates the 190-mer band that was excised for subsequent purification. 2: Purified 190-mer band excised from Lane 1. 3: Feasible answer sequences following sequential magnetic affinity purification to insure the presence of every node sequence. M: Molecular size reference using a 20-mer DNA ladder where the 100-mer was the brightest band (red arrow).

To test the validity of using ONPA to identify the optimal answer of a TSP, a linear programming (LP) model containing 5168 variables and 64 constraints was developed to "reverse engineer" the frequency of the feasible tours formed by DNA computing that would most closely match the individual ONP frequencies found in **Table 4** when broken down into their constituent ONPs. Since the data for each row in **Table 4** were obtained independently from the others, the model **1–3** was solved by inputting either the normalized and absolute ONP frequencies in **Table 4** as the F_{ij} parameters. In the normalized model, the F_{ij} were converted into percentages of each row total, whereas in the absolute model, the F_{ij} values were the total number of molecules detected (see **Table 5** footnotes a and b).

For the normalized model, the largest individual deviation is for the FG-ONP. The observed abundance F_{FG} was 0.98791. In the optimal solution, summed over all reverse-engineered tours ($\sum_{t=1}^{5040} h_{tFG}X_t$), arc FG was part of 0.99215 of the tours formed. Therefore, the FG-ONP would have a negative deviation $d_{FG}^- = 0.00425$ and a positive deviation $d_{FG}^+ = 0$. Many sets of positive and negative deviations exist that could equalize this constraint. However, the set that contributes the smallest amount to the objective results from the case in which one or both of the positive and negative deviations are zero, because the objective function adds both positive and negative deviations. In addition, there exist many possible solutions with X_t values such that $\sum_{t=1}^{5040} h_{tFG}X_t = F_{FG} = 0.98791$ exactly, but none of them were optimal because it would have increased the deviations for other ONPs. The same set of X_t values appeared in all 64 constraints and, while there is *no* set of X_t values that can satisfy all 64 instances of constraint **2** perfectly with no deviations, the model was designed to minimize the sum of the deviations for all ONPs simultaneously.

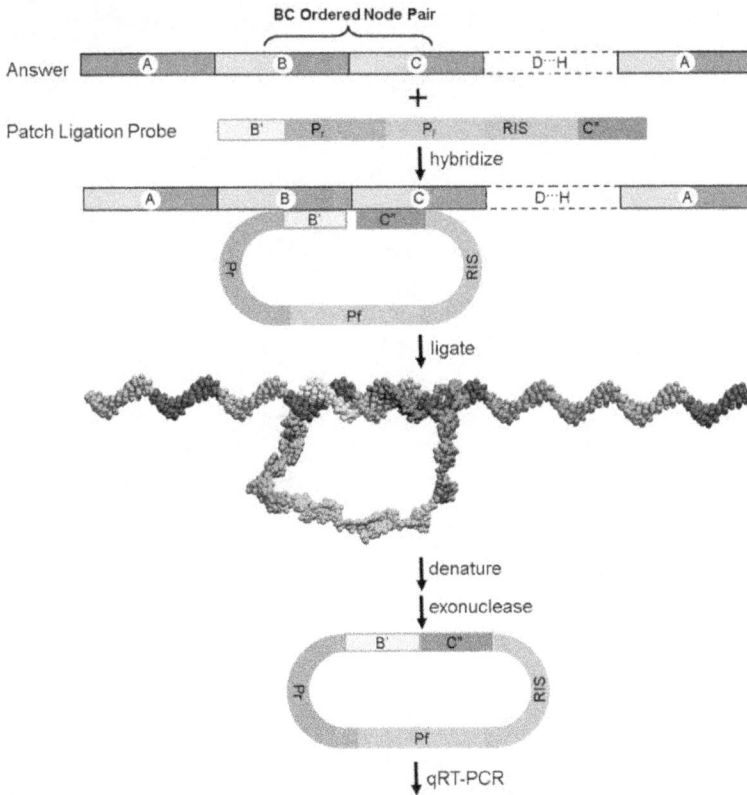

Figure 5.
Use of pair ligation probes (PLPs) to quantify the abundance of the nucleotide sequence encoding the BC ordered node pair (BC-ONP) where node B precedes node C. (1) The PLP hybridizes to 20 consecutive bases of the BC target sequence. (2) T4 ligase circularizes the hybridized PLP. (3) Circularized PLP is denatured from target, and Exonuclease I eliminated any nucleic acid not circularized. (4) qPCR quantifies PLP abundance upon addition of the TaqMan-MGB® reporter, PCR primers and nucleotides.

When the normalized ONPA data from **Table 4** were input, 99.01% of feasible tours formed by the DNA computer are estimated to have been the optimal tour ABCDEFGHA. Given the relative proportions of the ONPs formed, no other tour is likely to have comprised more than 0.38% of the tours formed (**Table 5**), which was the percentage for the second-most frequently formed tour in the reverse-engineering model ($A_SBFECDGHA_E$). Of the 5040 possible tours in the LP model, the optimal reverse-engineered set of tours had only 30 tours with a non-zero frequency. The objective function value in the model of the normalized data (the sum of the positive and negative deviations over all 64 ordered node pairs) was 0.02997.

Using the *absolute* molecule frequencies as the F_{ij} values, the LP model estimated that 339,742 molecules, or 99.13% of all feasible tour molecules, were the optimal answer. The next most frequently produced tour by the DNA computer was estimated to have formed only 1259 molecules (**Table 5**). As expected, the error levels increased dramatically when the model was used to reverse-engineer the tours formed from the absolute molecule frequency. This occurred because the row sums of molecules in **Table 4** were determined in separate qPCR measurements without controlling for the absolute amount of answer sequences used in the samples between rows. Had this been held constant during the measurement, the total

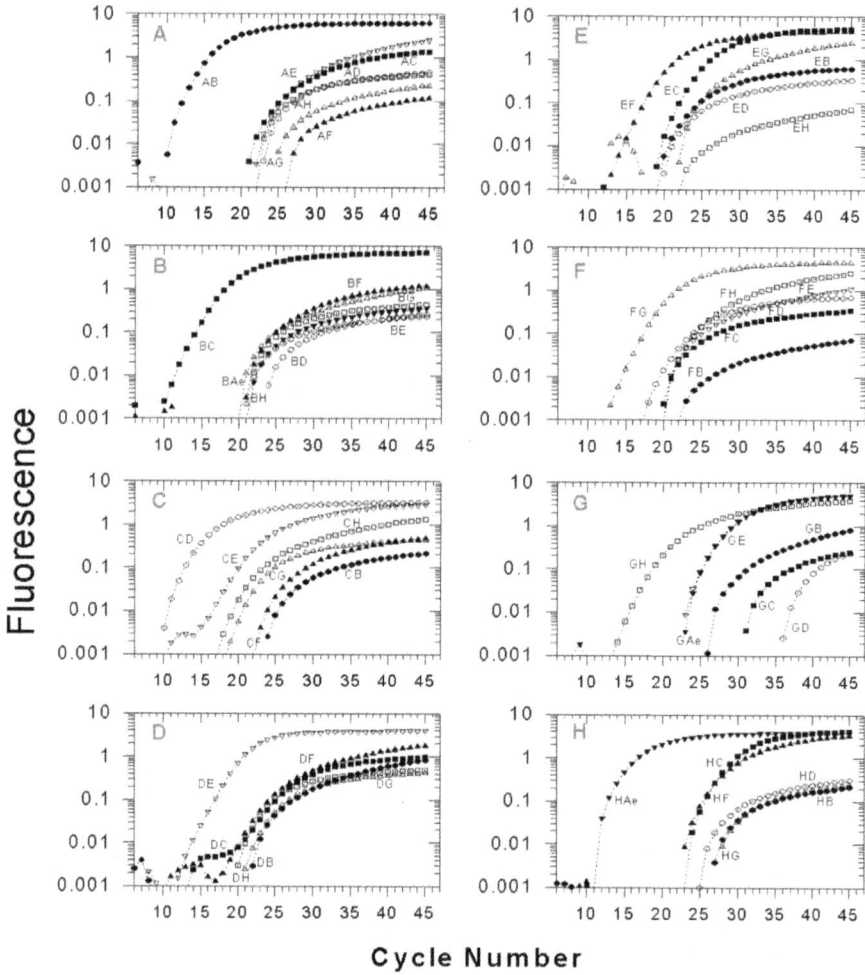

Figure 6.
Abundance of ONPs in the feasible answer sequences determined by PLP-qPCR. A separate PLP was designed to detect the amount of each of the 56 possible ONPs ($A_S B$, $A_S C$, etc.). Each PLP contained a qPCR reporter identification sequence for use in fluorescence detection with the TaqMan-MGB® reporter for NED. Plots of fluorescence intensity as a function of PCR cycle number are grouped by the preceding node in the ONP. Each of these groups was run as single-plex assays on a 96-well plate using aliquots from a common solution of answer sequences so that the assays within each group are comparable. The latter node in the ONP is indicated as B (●), C (■), D (○), E (▽), F (▲), G (△), H (□), and A_E (▼).

number of molecules in each row of **Table 4** should be the same because each feasible answer contains one copy of each node sequence. The total deviations summed to 9,057,230, which represents 81.57% of the total $\Sigma_i \Sigma_j F_{ij}$, compared with only 0.37% for the normalized model. It is noteworthy that, in the model based on the absolute data, the five ONPs with the greatest deviations from the observed frequencies (HA$_E$, A$_S$B, GH, BC, CD) were all part of the optimal tour, whereas only two such ONPs were among the top five deviations (FG, BF, DG, BC, and FD) when matching to the normalized ONPA. Despite the fact that the deviations derived from the absolute ONPs were much higher, the DNA computation still returned the optimal tour over 99% of the time.

A fully connected 8-city TSP was solved by the ordered node pair abundance (ONPA) approach through the use of a pair ligation probe quantitative real time

Prior node	Subsequent node							
	B	C	D	E	F	G	H	A_E
A_S	1.4×10^{6a}	10	3.5	2.9	0.019	0.19	1.4	*
B	*	4.8×10^5	0.49	6	10	9	4.9	21
C	0.04	*	2.2×10^5	420	0.18	6.4	22	0
D	28.5	190	*	3.5×10^5	395	320	48.5	750
E	640	1600	380	*	3.4×10^5	87	1.5	0
F	1.5	0	1800	1600	*	3.4×10^5	380	380
G	0.87	0.057	0	47	0	*	1.2×10^5	31
H	0.4	22	1.6	0	47	0.34	*	7.9×10^6

[a]All values are in 1000s of molecules.

Table 4.
ONPA in the set of feasible answers from the 8-city TSP determined by PLP-qPCR.

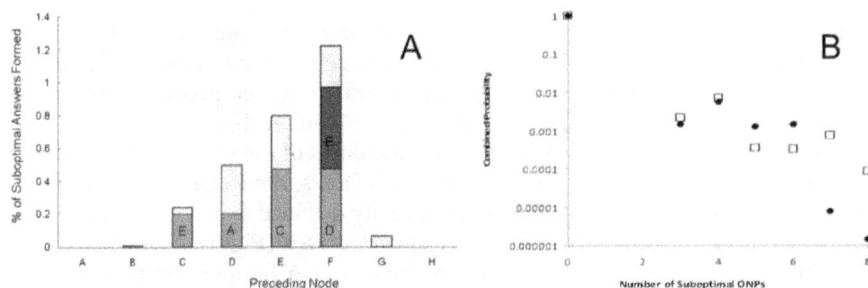

Figure 7.
*(A) Dominant suboptimal ONPs present in the answer sequences. The subsequent node in the ONP is indicated in each bar, and the sum of all other suboptimal ONPs in a given row from **Table 5** is indicated in yellow. (B) Combined probability of tours reverse engineered by the LP Model as a function of the Number of Suboptimal ONPs. The combined probability is the sum of the probabilities of all tours containing a given number of suboptimal ONPs generated from the normalized (●) and absolute data (□ of **Table 5**). The optimal tour did not contain suboptimal ONPs. From 1 to 9 tours were reverse engineered for each number of suboptimal ONPs shown.*

polymerase chain reaction (PLP-qPCR) system. The high specificity of the sequence-tagged hybridization and ligation that results from the use of PLPs significantly increased the accuracy of answer determination in DNA computing. When combined with the high throughput efficiency of qPCR, the time required to identify the optimal answer to the TSP was reduced from days to 25 min.

The reverse-engineering LP model provides additional evidence that the DNA computer can distinguish the quality of solutions to the TSP. Although the problem was set up artificially to have a single solution that was far better than any other possible solution, the reverse-engineering LP model confirms that the optimal solution was produced in far greater abundance than any other answer, and among the suboptimal solutions, those with greater optimality were generally produced in greater abundance than those that were less optimal.

Examination of the less frequently produced tours provides a measure of the sensitivity of DNA computer to the abundance of the arc sequence molecules initially input to carry out the computation. To make any change to the optimal tour $A_SBCDEFGHA_E$, at least three ONPs in the tour must be different. For example, the tour with penultimate optimality (**Figure 2e**) must have 3 ONPs that are different

Normalized ONPA F_{ij}[a]			Absolute ONPA F_{ij}[b]			
Top five tours[c]	# suboptimal ONPs	Probability	Top five tours[d]	# suboptimal ONPs	Molecules (in 1000s)	Probability
$A_sBCDEFGHA_E$	0	0.990131	$A_sBCDEFGHA_E$	0	339,742	0.991252
$A_sBFECDGHA_E$	4	0.003831	$A_sBFDECGHA_E$	4	1,259	0.003672
$A_sFDEBCGHA_E$	4	0.001860	$A_sFDEBCGHA_E$	4	512	0.001494
$A_sBCEDHGFA_E$	6	0.001099	$A_sBCEDFGHA_E$	3	258	0.000754
$A_sBCDFEGHA_E$	3	0.000733	$A_sBFECDGHA_E$	4	217	0.000634

[a]Based on ONPA as a percent of total molecules in each row in **Table 4** (i.e. ΣF_{ij} on Row A_s = 1.00, ΣF_{ij} on Row B = 1.00, etc.).
[b]Based on ONPA in **Table 5** (i.e. ΣF_{ij} on Row A_s = 1.5 e9, ΣF_{ij} on Row B = 4.5 e8, etc.).
[c]A total of 30 non-zero tours were reversed engineered from Normalized data with 0.0299702 total absolute deviations (objective function value).
[d]A total of 32 non-zero tours were reverse engineered from Absolute data with 9,057,230,000 total absolute deviations (objective function value).

Table 5.
Tour probabilities reverse engineered from observed ONPA by the LP model.

from the optimal (**Figure 2d**), while tours containing 4 different predominant ONPs (**Figure 2f**) would be even less optimal, and so forth. As shown in **Table 5**, among the 2nd-5th most frequently produced tours for models predicted from both the absolute and normalized data, all but one exhibit either three or four ONP changes from the optimal tour. When the probabilities of the tours were summed over all tours with a given number of suboptimal ONPs, the frequency of tours formed with more suboptimal ONPs exponentially declined as a function of an increase in the number of suboptimal ONPs incorporated (**Figure 7B**). Thus, not only did the DNA computer generate the optimal answer far more frequently, but other answers were generated in amounts that were inversely proportional to their objective function value.

Even though the data of **Table 4** were obtained in separate sets grouped according to each row without controlling for the abundance of molecules between rows (absolute data), the computation returned the correct answer, demonstrating the robustness of this approach. Several ONPs were not detected as products of the computation (e.g. FB). However, this does not imply that no molecules of these ONPs were formed. Although the amounts of these ONPs were below the level of detection, this did not alter the determination of the optimal answer to this problem because the efficiency of the tour that gave rise to the optimal answer was so much greater than any other tour. Identification of the optimal answers to some less obvious problems may require a more sensitive discrimination of ONPA. Under these circumstances, to quantify smaller amounts of ONPs or to discriminate between smaller differences in abundance of two specific ONPs, the PLP-qPCR can be repeated using larger aliquots of feasible answer solutions.

Acknowledgements

This work was supported by grants to WDF from DARPA-DSO and AFOSR (FA95500710219). The Fair Isaac Corp. (www.fico.com) provided the *Xpress* Optimization Suite to the university under their Academic Partnership Program.

Author details

Fusheng Xiong[1], Michael Kuby[2] and Wayne D. Frasch[1*]

1 School of Life Sciences, Arizona State University, Tempe, AZ, USA

2 School of Geographical Sciences and Urban Planning, Arizona State University, Tempe, AZ, USA

*Address all correspondence to: frasch@asu.edu

IntechOpen

References

[1] Lin S. Computer solutions of traveling salesman problem. Bell System Technical Journal. 1965;**44**:2245

[2] Rosenkrantz DJ, Stearns RE, Lewis PM. An analysis of several heuristics for the traveling salesman problem. SIAM Journal of Computing. 1977;**6**:563-581

[3] Crowder H, Padberg MW. Solving large-scale symmetric traveling salesman problems to optimality. Management Science. 1980; **26**:495-509

[4] Jünger M, Reinelt G, Rinaldi G. The traveling salesman problem. In: Ball MO, Monma CL, Nemhauser GL, editors. Handbooks in Operations Research and Management Science. Amsterdam: Elsevier Science B.V; 1995. pp. 225-330

[5] Ryu H. A revisiting method using a covariance traveling salesman problem algorithm for landmark-based simultaneous localization and mapping. Sensors. 2019;**19**:E4910

[6] Miao K, Duan H, Qian F, Dong Y. A one-commodity pickup-and-delivery traveling salesman problem solved by a two-stage method: A sensor relocation application. PLoS One. 2019;**14**: e0215107

[7] Kahng AB, Reda S. Match twice and stitch: A new TSP tour construction heuristic. Operations Research Letters. 2004;**32**:499-509

[8] Adleman LM. Molecular computation of solutions to combinational problems. Science. 1994; **266**:1021-1024

[9] Lipton RJ. DNA solution of hard computational problems. Science. 1995; **268**:542-545

[10] Lee JY, Shin SY, Park TH, Zhang BT. Solving traveling salesman problems with DNA molecules encoding numerical values. Biosystems. 2004;**78**: 39-47

[11] Spetzler D, Ziong F, Frasch WD. Heuristic solution to a 10-city Asymmetric traveling salesman problem using probabilistic DNA Computing. Lecture Notes in Computer Science. 2008;**4848**:152-160

[12] Xiong FS, Spetzler D, Frasch WD. Solving the fully-connected 15-city TSP using probabilistic DNA computing. Integrative Biology. 2009;**1**:275-280

[13] Sharma D, Ramteke M. A note on short-term scheduling of multi-grade polymer plant using DNA computing. Chemical Engineering Research and Design. 2018;**135**:78-93

[14] Xu F, Wu T, Shi X, Pan L. A study on a special DNA nanotube assembled from two single-stranded tiles. Nanotechnology. 2019;**30**(115602):1-6

[15] Woods D, Doty D, Myhrvold C, Hui J, Zhou F, Yin P, et al. Diverse and robust molecular algorithms using reprogrammable DNA self-assembly. Nature. 2019;**567**:366-372

[16] Yamamoto M, Kameda A, Matsuura M, Shiba T, Kawazoe Y, Ohuchi A. A separation method for DNA computing based on concentration control. New Generation Computing. 2002;**20**:251-261

[17] Lee J-Y, Shin S-Y, Augh SJ, Park TH, Zhang B-T. Temperature gradient-based DNA computing for graph problems with weighted edges. Lecture notes in Computer Science. 2003;**2568**:73-84

[18] Yamamura M, Hiroto Y, Matoba T. Solutions of shortest path problems by

concentration control. Lecture Notes in Computer Science. 2002;**2340**:231-240

[19] Henco K, Harders J, Wiese U, Riesner D. Temperature gradient gel electrophoresis (TGGE) for the detection of polymorphic DNA and RNA. Methods in Molecular Biology. 1994;**31**:211-228

[20] Riesner D, Steger G, Wiese U, Wulfert M, Heibey M, Henco K. Temperature-gradient gel electrophoresis for the detection of polymorphic DNA and for quantitative polymerase chain reaction. Electrophoresis. 1992;**13**:632-636

[21] Ibrahim Z, Rose JA, Suyama A, Khalid M. Experimental Implementation and analysis of a DNA computing readout method based on real-time PCR with TaqMan probes. Natural Computing. 2008;**7**:277-286

[22] Szemes M, Bonants P, de Weerdt M, Baner J, Landegren U, Schoen CD. Diagnostic application of padlock probes-multiple detection of plant pathogens using universal microarrays. Nucleic Acids Research. 2005;**33**:e70

[23] Xiong F, Frasch WD. Padlock probe-mediated qRT-PCR for DNA computing answer determination. Natural Computing. 2010;**10**:947-959

[24] Brüggemann W. A minimal cost function method for optimizing the age-depth relation of deep-sea sediment cores. Paleoceanography. 1992;7: 467-487

[25] Kuby MJ, Cerveny RS, Dorn RI. A new approach to paleoclimatic research using linear programming. Palaeogeography Palaeoclimatocology Palaeoecology. 1997;**129**:251-267

[26] Leinen M, Pisias N. An objective technique for determining end-member compositions and for partitioning

sediments according to their sources. Geochimica et Cosmochimica Acta. 1984;**48**:47-62

[27] Narula SC, Wellington JF. Selection of variables in linear-regression using the minimum sum of weighted absolute errors Criterion. Technometrics. 1979; **21**:299-306

Chapter 5

Comparative Study of Algorithms Metaheuristics Based Applied to the Solution of the Capacitated Vehicle Routing Problem

Fernando Francisco Sandoya Sánchez,

Carmen Andrea Letamendi Lazo

and Fanny Yamel Sanabria Quiñónez

Abstract

This chapter presents the best-known heuristics and metaheuristics that are applied to solve the capacitated vehicle routing problem (CVRP), which is the generalization of the TSP, in which the nodes are visited by more than one route. To find out which algorithm obtains better results, there are 30 test instances used, which are grouped into 3 sets of problems according to the position of the nodes. The study begins with an economic impact analysis of the transportation sector in companies, which represents up to 20% of the final cost of the product. This case study focuses on the CVRP for its acronym capacitated vehicle routing problem, analyzing the best-known heuristics such as Clarke & Wright and sweep, and the algorithms GRASP and simulated annealing metaheuristics based.

Keywords: vehicle routing problem, VRP, CVRP, heuristics, metaheuristics, Clarke & Wright, sweep heuristic, GRASP, simulated annealing

1. Introduction

Logistics as a science has its origins in the military area; the transportation of weapons, food, and men at the service was coordinated through it. With the passage of time, the concept began to be applied in the business field, and for a long period of time, the logistics function was considered as a habitual, operational, and necessary activity to take the products from the seller to the buyer [1]. A little later, starting in the 1950s, a cycle of growth and constant demand increase was experienced throughout the world, which caused the production and sales capacity to exceed the companies' ability to distribute products. Thus, in those years, delivering orders on time became a problem due to poor compliance. Then, in 1980, the concept of response time was created, which is the union between the concept of physical distribution and material management; specialists realized that the faster the response time to the customer, the more the profitability of the company increased.

As the concepts were changing, the methods as well and the companies looked for ways to become efficient; in this way they expanded the activities related to logistics and determined that one of the heaviest items is transportation costs, representing on average, between 10% and 20% of the final cost of the product or service [2].

Although transport decisions are expressed in a variety of ways, the main ones are mode selection, route design, vehicle programming, and shipment consolidation [1]. In relation to the route design problem, the problem is commonly known as vehicle routing problem (VRP). Both the companies that own the transport service as part of their processes and the companies that provide the service seek to optimize resources within the route selection, since a good selection brings savings in time, resources such as fuel, maintenance of the fleet, salaries, and improvements, among others, in service indicators as a promise of product delivery.

The VRP can be considered as the natural extension of the TSP, in the sense that unlike the TSP, in the VRP we consider that the vehicles, or the agents in charge of providing a service to the nodes, have a limited capacity; therefore, most likely the entire route cannot be made through a single route, with a single vehicle that leaves and returns to the storage, traveling all the nodes, but to respect the restriction of the limited capacity of the vehicles so. In general, several routes are required, or what is the same, the solution of the VRP will be a set of Hamiltonian cycles that start from the deposit and such that each node is traveled only once.

2. Vehicle routing problem

The vehicle routing problem (VRP) consists in determining a set of routes for a fleet of vehicles that depart from one or more warehouses to meet the demand of several geographically dispersed customers [3].

The VRP objective is to meet the demand of the customers, optimizing some objective, which is generally the total cost involved in the routes, which is affected by the vehicular congestion of large cities, the high-energy consumption of cargo vehicles, and other factors.

Since the VRP problem is a generalization of the TSP, and knowing that the TSP is of the NP-hard problem class [4], it is concluded that the VRP is also a difficult problem of the NP-hard class.

The VRP model has many classifications by the different characteristics that can be included or considered in it. The most basic version is reflected with the CVRP capability (for the acronym of capacitated vehicle routing problem). The CVRP has the following assumptions:

The fleet of vehicles is homogeneous, that is to say all cargo vehicles have the same characteristics:

I. The demand is known in advance, that is, the quantity to be delivered for each client is known; this means that the demand is deterministic.

II. Each vehicle will carry the entire delivery to customers, prohibiting the distribution of fractional or partial loads that would later be completed by another vehicle.

III. All vehicles in the fleet have exactly the same load capacity.

IV. The starting point of the vehicles is only one and is considered a central warehouse.

V. Vehicles have capacity restrictions that are known in advance.

3. Classic heuristics to solve VRP

Heuristics are simple processes that perform a limited space search and generate acceptable solutions in moderate calculation times; an important characteristic of these methods is that they are designed to solve a specific optimization problem, and in general they cannot be used to solve other optimization problems. A more advanced class are the so-called metaheuristics, which are considered more advanced methods than heuristics, in the sense that they guide their construction and, therefore, are general purpose [5].

There are many advantages, and also disadvantages, when using heuristic algorithm methods to solve optimization problems, as described [6] within the reasons to use heuristic methods which are as follows:

I. The problem is that no exact method for its resolution is known.

II. Although there is an exact method to solve the problem, its execution is computationally very expensive.

III. A heuristic method is more flexible than an exact method, that is, difficult modeling conditions can be incorporated.

IV. The heuristic method is used as part of a global process that certifies an optimal solution. There are two possibilities:

a. The heuristic method provides a good initial starting solution.

b. The heuristic method participates in an intermediate step of the procedure, such as the selection rules of the variable to enter the base in the simplex method.

3.1 Savings based heuristics

There are several types of heuristic methods to solve the VRP, which are addressed extensively in Braekers et al. [3], trying to generate broad, nonexclusive categories, where the best-known heuristics are located to solve this problem; among them one of the most used and popular algorithms is the one of Clarke & Wright, and that has had contributions from different authors [7].

This algorithm is based on successively combining subtours until a Hamiltonian cycle is obtained, of which the subtours have a common node or vertex called base or initial.

The method can be described as follows:

- Having a solution of two different routes $(0, ..., i, 0)$ and $(0, j, ..., 0)$ can be joined by creating a new route $(0, ..., i, j, ..., 0)$.

- The distance savings obtained by the union is

$$s_{ij} = c_{i0} + c_{0j} - c_{ij} \tag{1}$$

In Eq. (1) s_{ij} is the savings on the total distance traveled if the two routes $(0, ..., i, 0)$ and $(0, j, ..., 0)$ are joined.

- An initial solution is started in this algorithm, and the unions that give greater savings are made as long as they do not violate the restrictions of the problem.

- When the maximum saving is negative, the combinations of the routes will increase the distance traveled, but the amount of routes in the solution will decrease; depending on the characteristics of each problem, this can generate circular or radial routes that can be avoided by placing a reference value λ, which penalizes the union routes with distant customers. Saving is proposed as

$$s_{ij} = c_{i0} + c_{0j} - \lambda c_{ij} \tag{2}$$

In Eq. (2) s_{ij} represents penalized savings with the weight λ in the total distance traveled if the two routes $(0, \dots, i, 0)$ and $(0, j, \dots, 0)$ are joined, which prevents when possible, merging routes with nodes far apart.

3.1.1 Application of savings algorithm

Step 1: With the coordinates of each client or city, prepare the distance matrix.
Step 2: Calculate the s_{ij} savings table for each pair of nodes.
Step 3: For each client or city i, build the route $(0, i, 0)$.
Step 4: Order savings from highest to lowest.
Step 5: Starting with the greatest savings, join the corresponding nodes, so that $s_{ij} = max\ s_{ij}$, where the maximum is taken between the savings that have not yet been considered; the route, $r_{i^* j^*}$, will be created, if i^* is the last customer of de r_{i^*}; and j^* is the first customer of r_{j^*}. Remove $s_{i^* j^*}$ from future considerations. Repeat step 5 until there are no more combinations of savings.

3.1.2 Example of application of the savings algorithm

A company wants to solve the problem of routing and design of the fleet of its product x to its 10 customers in the city and has a homogeneous fleet of trucks with capacity for 100 units of x product, with locations and demand shown in **Table 1**.
There are details in the Cartesian coordinates of the warehouse and each customer with the demand, while in **Figure 1**, the position of each customer and the warehouse is shown.

Step 1: The matrix of Euclidean distances between each pair of nodes is calculated: **Table 2** shows the distance matrix between all customers along with the warehouse. This matrix is symmetrical, that is, it has the same distance to go from client i to client j and vice versa. Point 0 has been considered for the warehouse (whs).
Step 2: Once the distance matrix is obtained, the savings are calculated. For the savings matrix, no row or column is placed for the warehouse.
For example, the savings between customer 1 and customer 2 is

$$s_{12} = c_{1bdg} + c_{bdg2} - c_{12} \tag{3}$$
$$s_{12} = 25.46 + 19.80 - 5.66 = 39 \tag{4}$$

In **Table 3** all the savings are shown; in the same way the matrix is symmetric.
Step 3: The route $(0, i, 0)$ is built for each client. In **Figure 2** each route is shown from the warehouse to each customer and back to the warehouse.
Step 4: Savings are organized from the highest to lowest.

customer	coordinates		demand
	x	*y*	
1	21	21	17
2	25	25	25
3	20	44	10
4	44	34	29
5	54	19	20
6	54	44	15
7	25	58	28
8	54	64	36
9	18	27	14
10	38	50	24
whs	39	39	

Table 1.
Cardinal coordinates of storage and customers with their demand.

In the list of savings to choose, only the savings that can be chosen are considered.

When the list is prepared with all the savings, those savings that one or both clients have already considered in a previous route are discarded.

Step 5: To assemble the routes, the restrictions are considered; for this example the only restriction is the capacity of the truck that does not exceed 100 units of the product x.

For the first route, the highest savings are chosen and placed in the form $(0, i, 0)$; in this case 0 is the warehouse (whs), as the savings are chosen to add them to the existing route or create a new one; the demands of each client are added, and the route is closed when the sum of the demands is equal to or less than the capacity of the truck 100 units, but when adding one more client, the demand exceeds the capacity, and you can no longer choose that customer.

The composition of the routes is displayed step by step in **Figure 3**, and the complete route diagram is shown in **Figure 4**. Below is the composition of the routes with the demands.

Route 1: whs, c1, c9, whs	Route demand 1: 17 + 14 = 31
Route 1: whs, c2, c1, c9, whs	Route demand 1: 31 + 25 = 56
Route 2: whs, c3, c7, whs	Route demand 2: 10 + 28 = 38
Route 1: whs, c2, c1, c9, c3, c7, whs	Route demand 1: 56 + 38 = 94
Route 2: whs, c6, c8, whs	Route demand 2: 15 + 36 = 51
Route 2: whs, c6, c8, c10, whs	Route demand 2: 51 + 24 = 75

Route 2: whs, c5, c6, c8, c10, whs	Route demand 2: 75 + 20 = 95
Route 3: whs, c4, whs	Route demand 3: 29

Figure 1.
Customer and warehouse positioning.

Cij	whs	c1	c2	c3	c4	c5	c6	c7	c8	c9	c10
whs	—	25.46	19.80	19.65	7.07	25.00	15.81	23.60	29.15	24.19	11.05
c1	25.46	—	5.66	23.02	26.42	33.06	40.22	37.22	54.20	6.71	33.62
c2	19.80	5.66	—	19.65	21.02	29.61	34.67	33.00	48.60	7.28	28.18
c3	19.65	23.02	19.65	—	26.00	42.20	34.00	14.87	39.45	17.12	18.97
c4	7.07	26.42	21.02	26.00	—	18.03	14.14	30.61	31.62	26.93	17.09
c5	25.00	33.06	29.61	42.20	18.03	—	25.00	48.60	45.00	36.88	34.89
c6	15.81	40.22	34.67	34.00	14.14	25.00	—	32.20	20.00	39.81	17.09
c7	23.60	37.22	33.00	14.87	30.61	48.60	32.20	—	29.61	31.78	15.26
c8	29.15	54.20	48.60	39.45	31.62	45.00	20.00	29.61	—	51.62	21.26
c9	24.19	6.71	7.28	17.12	26.93	36.88	39.81	31.78	51.62	—	30.48
c10	11.05	33.62	28.18	18.97	17.09	34.89	17.09	15.26	21.26	30.48	—

Table 2.
Distance matrix.

S_{ij}	c1	c2	c3	c4	c5	c6	c7	c8	c9	c10
c1	—	39.60	22.08	6.11	17.40	1.04	11.84	0.41	42.93	2.89
c2		—	19.80	5.85	15.18	0.94	10.40	0.35	36.71	2.67
c3			—	0.72	2.44	1.46	28.38	9.36	26.72	11.72
c4				—	14.04	8.74	0.06	4.60	4.33	1.03
c5					—	15.81	0.00	9.15	12.31	1.16
c6						—	7.21	24.97	0.19	9.77
c7							—	23.14	16.01	19.38
c8								—	1.72	18.94
c9									—	4.75
c10										—

Table 3.
Savings matrix.

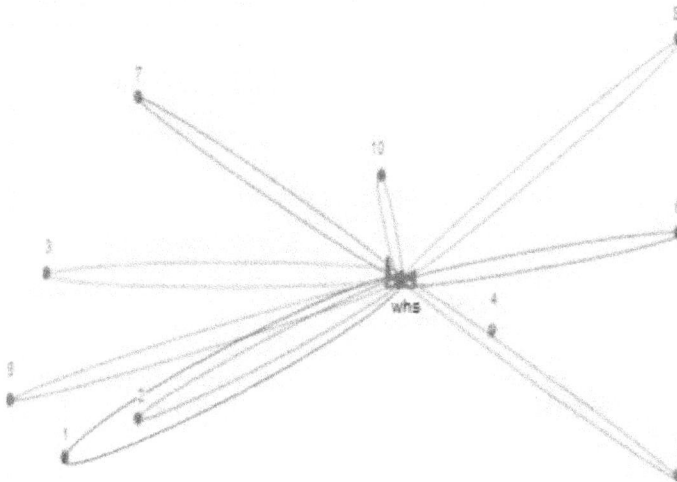

Figure 2.
Clarke & Wright heuristics step 1 route development (o, i, o).

Consequently, Clarke & Wright algorithm determines a solution for the routing problem in which the distance traveled is 204.20 units in length.

3.2 Heuristic method of assigning first, routing after

Sweep heuristics are the best-known method of assigning first, routing later.

This method is solved in two phases. First, groups of customers called clusters are created considering the capacity constraints of the vehicles, and second for each cluster, a route is generated that visits all customers.

In sweep heuristic, clusters are created by turning a half-straight in the central tank from the horizontal counterclockwise; after that the customers are incorporated into the mentioned group until the maximum capacity restriction of the vehicles is met.

This heuristic is used to find solutions to geographical problems, that is to say in which the nodes or vertices correspond to a point in the plane. It is assumed that the

Route 1: whs, c1, c9, whs	Route 1: whs, c2, c1, c9, whs

Route 2: whs, c3, c7, whs	Route 1: whs, c2, c1, c9, c3, c7, whs

Route 2: whs, c6, c8, whs	Route 2: whs, c6, c8, c10, whs

Route 2: whs, c5, c6, c8, c10, whs	Route 3: whs, c4, whs

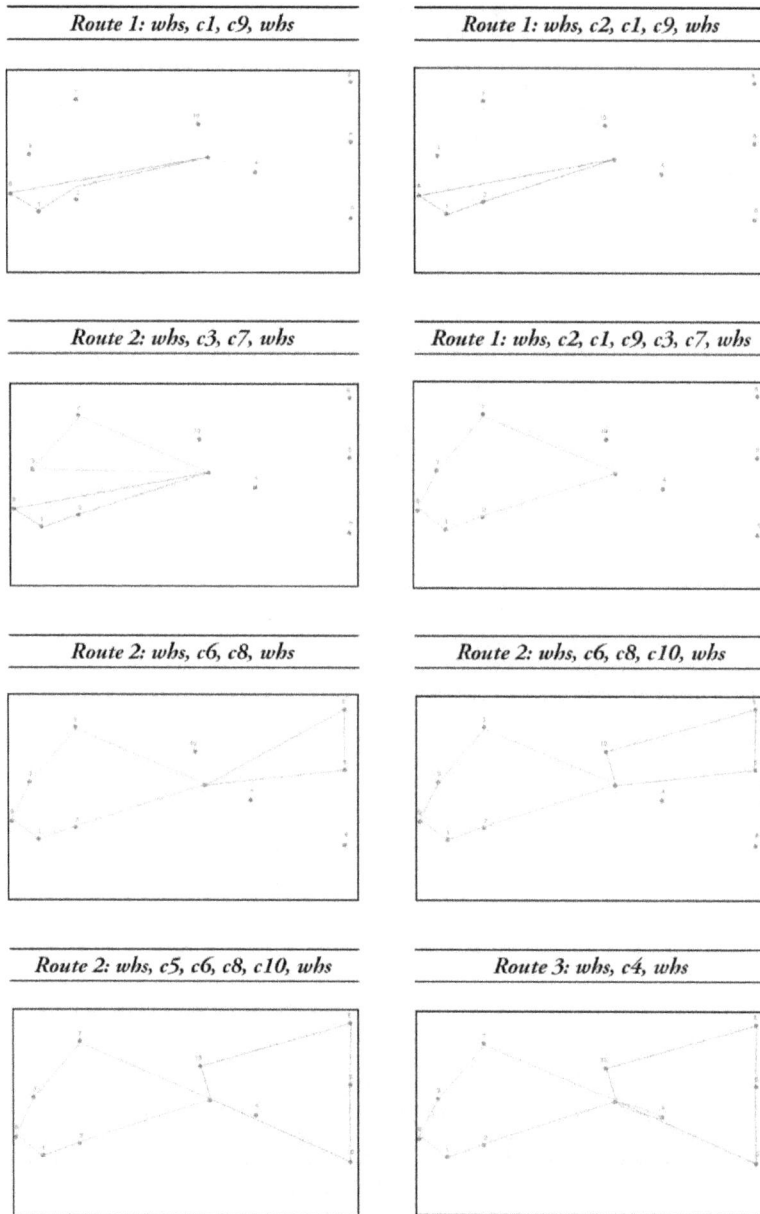

Figure 3.
Step by step: routing of Clarke & Wright heuristics.

location of each client i can be represented through its polar coordinates (r_i, θ_i) having a single central deposit. It defines

$$\theta_i = \arctan \left[\frac{Y(i) - Y(whs)}{X(i) - X(whs)} \right]$$

$$where \quad -\pi < r_i < 0 \quad si \quad Y(i) - Y(whs) < 0$$

$$y \quad 0 < r_i < \pi \quad si \quad Y(i) - Y(1) \geq 0, (i = 1, 2, \dots, n)$$

$$r_i = polar \ radio \ coordinate \ of \ the \ i \ th \ position \ (i = 1, 2, \dots, n)$$

(5)

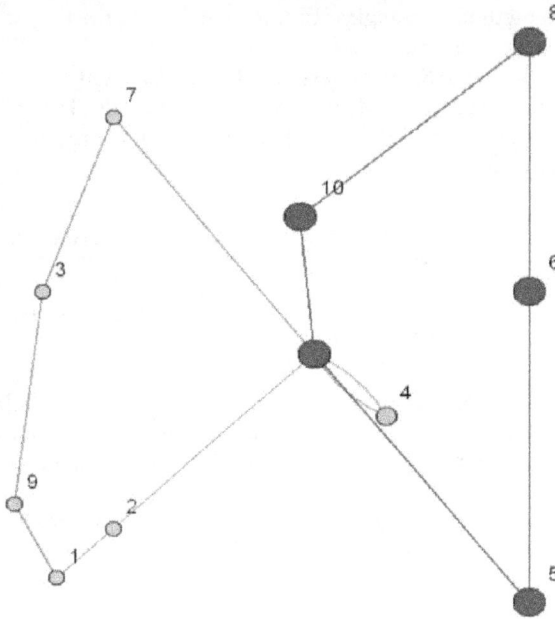

Figure 4.
Final routes by Clarke & Wright algorithm.

3.2.1 Steps for sweep heuristics

Step 1: Prepare the table of the location of the nodes in polar coordinates.

Step 2: Customers or cities are sorted in ascending order by θ; if two clients or cities have the same θ value, the one with the lowest r value is chosen first. Then a customer or city w is selected to start and make $k = 1$ y $C_k = \{w\}$.

Step 3: If all clients or cities are in a cluster, go to step 4. Otherwise, a client or city is selected; w_i and w_i are added to C_k if you do not exceed the capacity restrictions; if you exceed them, create a new cluster for which, $k = k + 1$ and $C_k = \{w_i\}$. Repeat step 2 until there are no clients or cities without a cluster.

Step 4: For each cluster C_k for $t = 1, \dots, k$, solve traveling salesman problem (TSP) with its clients and a solution that can be a local optimum is obtained until not checking otherwise.

Step 5: Return to step 2 to reorder customers where the first becomes the last, the second the first, and so on until the original sorting. For each change, steps 3 and 4 are performed again, and the best of the solutions obtained is taken.

3.2.2 Example of sweep heuristics

We will take the example of the savings algorithm.

Step 1: Formula (5) is used to obtain the polar coordinate table, where θ_i is expressed in radians and r_i is the directed distance. **Table 4** shows the polar coordinates for each client i. The change of polar coordinates is displayed (**Figure 5**).

Step 2: It is sorted by θ_i from least to greatest, as seen in **Table 5**.

Step 3: To elaborate the routes, it is done in two phases, the first one where the clients are grouped by the sweep method and the second one where a TSP is resolved (step 4).

For the sweep method, the angles from the smallest to the largest are chosen, and it moves counterclockwise.

As can be seen in **Table 5**, customers are already sorted in ascending order by their angular polar coordinate, and customers are chosen until they fail to comply with the capacity restriction of the truck that is 100 units of product x. Considering this the routes are as follows:

Route 1: whs, c9, c2, c1, c5, whs	Route demand 1: 14+25+17+20 = 76
Route 2: whs, c4, c6, c8, whs	Route demand 2: 29+15+36 = 80
Route 3: whs, c10, c7, c3, whs	Route demand 3: 24+28+10 = 62

In **Figure 6**, the sweeps are visualized starting with client 9 that has the greatest angle, thus grouping them in zones in this case by colors and within each one for their best distance. The sweep groups customers do not violate the restriction of the truck.

In **Figure 7**, the solution is shown with three routes before the TSP is applied.

Step 4: In the second phase to each route already generated in the first, it is resolved by TSP, for this case with the nearest node.

	r_i	θ_i
c1	25.46	−2.36
c2	19.80	−2.36
c3	19.65	2.88
c4	7.07	−0.79
c5	25.00	−0.93
c6	15.81	0.32
c7	23.60	2.21
c8	29.15	1.03
c9	24.19	−2.62
c10	11.05	1.66

Table 4.
Polar coordinates for each i.

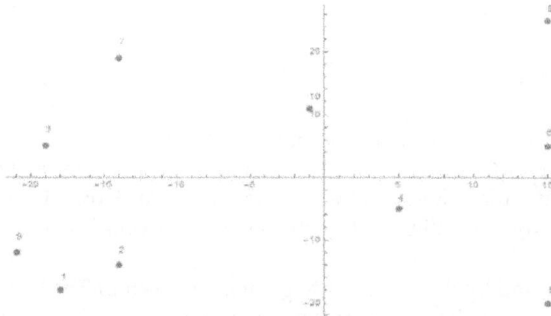

Figure 5.
Node location.

	r_i	θ_i
c9	24.19	−2.62
c2	19.80	−2.36
c1	25.46	−2.36
c5	25.00	−0.93
c4	7.07	−0.79
c6	15.81	0.32
c8	29.15	1.03
c10	11.05	1.66
c7	23.60	2.21
c3	19.65	2.88

Table 5.
Ascending ordering of customers.

The routes are as follows:

Route 1: whs, c9, c1, c2, c5, whs	Route demand 1: 14+17+25+20 = 76
Route 2: whs, c4, c6, c8, whs	Route demand 2: 29+15+36 = 80
Route 3: whs, c10, c7, c3, whs	Route demand 3: 24+28+10 = 62

Thus, the sweep algorithm has a local solution for the routing problem in which the distance traveled is 222.36 units in length.

In **Figure 8**, the route diagram is displayed.

Step 5: Repeat step 2 where the customers already ordered from **Table 5**, continue to rotate the position until the first returns to be first, and for each rotation, steps 3 and 4 are performed, and after all iterations, the best is selected.

Below is the iteration that had the best result. **Table 6** shows the fifth iteration of nine where you start with client six (**Figure 9**).

After performing step 3 in **Figure 10**, the grouping of customers is appreciated to not violate the restriction of the truck's capacity.

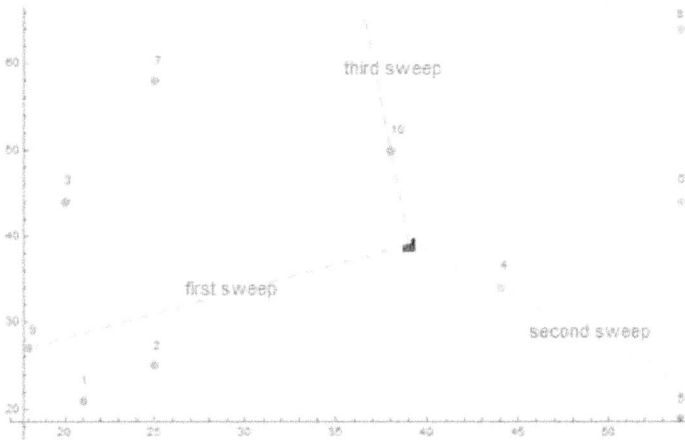

Figure 6.
First phase of sweep heuristics, grouping.

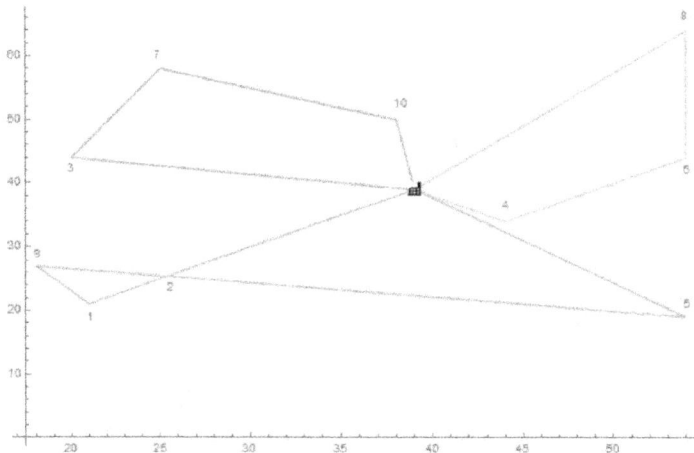

Figure 7.
First phase of sweep heuristics, solution with three routes.

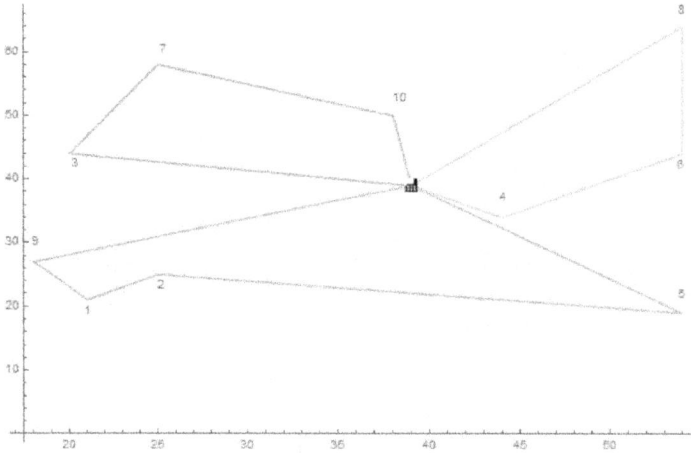

Figure 8.
Routes by sweep algorithm.

	r_i	θ_i
c6	15.81	0.32
c8	29.15	1.03
c10	11.05	1.66
c7	23.60	2.21
c3	19.65	2.88
c9	24.19	−2.62
c1	25.46	−2.36
c2	19.80	−2.36
c5	25.00	−0.93
c4	7.07	−0.79

Table 6.
Customer ordering—fifth iteration.

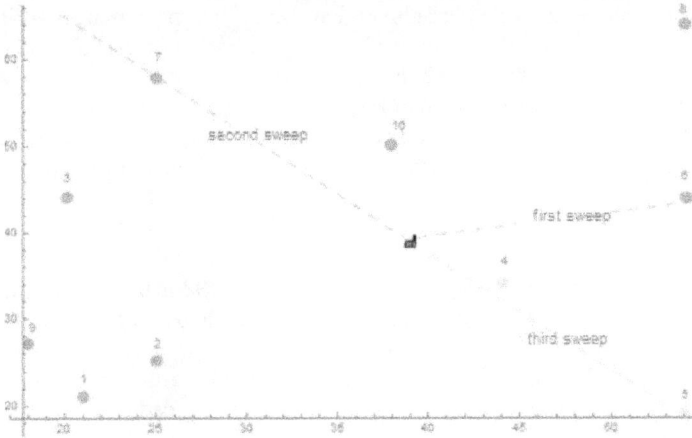

Figure 9.
First phase of sweep heuristics, grouping—fifth iteration.

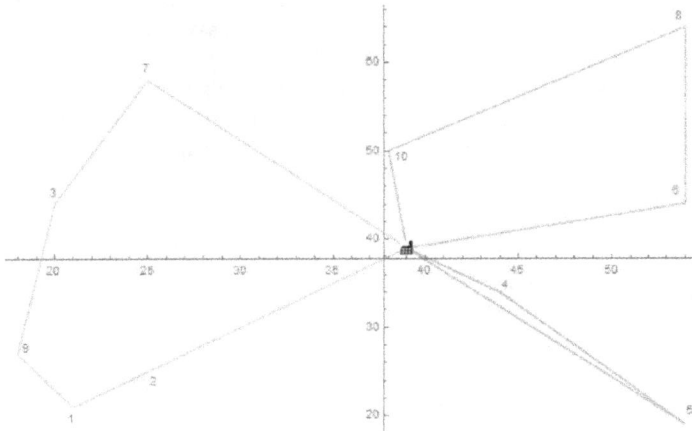

Figure 10.
Routes by sweep algorithm—fifth iteration.

In step 4, each grouping is resolved with a TSP, and the following routes are obtained:

Route 1: whs, c6, c8, c10, whs	Route demand 1: 15+36+24 = 75
Route 2: whs, c7, c3, c9, c1, c2, whs	Route demand 2: 28+10+14+25+17 = 94
Route 3: whs, c5, c4, whs	Route demand 3: 20+29 = 49

In **Figure 10**, the diagram of the routes of the fifth iteration is displayed, which obtained the best response.

The sweep algorithm determines a solution for the routing problem in which the distance traveled is 205.96 units in length, that is, a solution of lower quality than that obtained by the Clarke & Wright algorithm with 204.20 units in length.

4. Metaheuristics

The term metaheuristics first appeared in the seminal article about taboo search (Glover, 1987). The term metaheuristics is obtained by putting the suffix "meta" before the word heuristic, which means "beyond" or "at a higher level."

Metaheuristics are generic procedures that, through approximate algorithms, guide a subordinate heuristic by combining the exploration of the solution space for optimization problems, obtaining better results than classical heuristics in a longer period; however, this time is less than if the exact methods are used.

Metaheuristics that have been considered for this comparative study are shown below, which correspond to constructive and local search procedures [6].

4.1 GRASP

GRASP methods had their origins at the end of the 1980s in order to find a solution to problems of joint coverings, and in 1995 by Feo and Resende, this metaheuristic is of general purpose [8].

The word GRASP comes from the acronym of greedy randomized adaptive search procedures that would be something like search procedures based on voracious adaptive random functions.

GRASP has a multistart process in which each step has a construction and an improvement phase. In the construction phase, the constructive heuristic process obtains a good initial solution, which is improved in the second phase by a local search algorithm. The best of all solutions examined is saved as the final result.

There are many implementations of GRASP metaheuristics, including variants and hybridizations with other procedures such as variable neighborhood search or path relinking, with which this metaheuristic has proven to work very well in practice as demonstrated in Marti and Sandoya [9]. A simple scheme to represent the operation of this algorithm is as follows:

While (stop condition)
Construction phase:

- Choose a list of candidate elements.

- Have a restricted list with the best candidates.

- Select an item randomly from the restricted list.

Improvement phase:

- Perform a local search process based on the solution built until it can no longer be improved.

Update:

- If the solution obtained improves to the best stored, update it.

In the construction phase, a possible solution is built iteratively, considering an element in each step. In each iteration the choice of the next element to be added to the partial solution is determined by a greedy function, which examines the benefit of adding each of the elements according to the objective function and choosing the best one.

This metaheuristic works with a restricted list of the best candidates, which makes the best candidate randomly selected for each iteration of the construction phase.

In the improvement phase, the results that are obtained from the construction phase are not usually local optimal; therefore, a local search procedure is applied as post-processing to perfect the solution obtained.

Performing several iterations is a way of sampling the solution space.

4.2 Simulated annealing

The simulated annealing metaheuristics was introduced in the 1950s by Metropolis Hastings to be used in the field of statistical thermodynamics simulating cooling processes of a material.

In 1983 the method was refocused to solve combinatorial optimization problems of great complexity by Scott Kirkpatrick, C. Daniel Gelatt and Mario P. Vecchi, and independently in 1985 by Vlado Cerny. For its implementation ease, this metaheuristic had a great boom in the 1980s.

Simulated annealing has its procedure based on local search by environments that is characterized by an acceptance criterion of neighboring solutions that are adapted throughout its execution.

A temperature variable is used, T, that determines the extent to which neighboring solutions, worse than the current n, can be accepted. This variable is about starting it with a high value, which is called the initial temperature, T_0, which generates a high probability of accepting a nonimprovement movement. In each iteration the temperature decreases through a temperature cooling mechanism, α, having a smaller probability until approaching the optimal solution and reaching a final temperature, T_f. Costs also decrease as the temperature decreases, making it increasingly difficult to accept bad movements in search of the solution.

In each iteration a specific number of neighbors is generated, which can be fixed for the entire execution or depend on each iteration.

Each time a neighbor is generated, the acceptance criterion is applied to see if it replaces the current solution:

- If the neighboring solution is better than the current one, it is automatically accepted, as it would be done in the classic local search.

- If the neighboring solution is worse than the current one, there is still a chance that the neighbor will replace the current solution. This allows the algorithm to exit from local optimum, in which the classic local search would be trapped.

This model is given by the following structure:

Take an initial solution x
Take an initial temperature T
While (not frozen)
Perform L times
Take x' from $N(x)$
$d = f(x') - f(x)$
If $(d < 0)$ do $x = x'$
If $(d > 0)$ do $x = x'$ with $p = e^{-d/T}$
Take action of the cooling mechanism $(T = rT)$

The following parameters are determined:

I. Initial temperature: it is established by doing a series of tests to reach a certain fraction of accepted movements.

II. Cooling speed r.

III. Length L that must be proportional to the expected size of $N(x)$.

IV. When the cooling sequence ends, it is frozen $cont = cont + 1$ when a temperature is completed and the percentage of movements accepted is less than $MinPercent.cont = 0$ when the best stored solution is improved.

5. Implementation of heuristics and metaheuristics for the resolution of CVRP

5.1 Test instances

The cases to be evaluated are divided into three groups classified by the type of client with 10 examples each. Next, some tables will be presented, which have the name of instance, the truck's capacity in column cap, the number of customers in column n, the number of vehicles to be used in column k, and the optimal solution in column opt.

I. Clustered clients, as shown in **Table 7**, belong to the Augerat B set in 1995 [10] and specify that the coordinates are points between $[0,100] \times [0,100]$ in the grid that are chosen to form neighborhood groups (NC) closest, where $k \leq NC - 1$. The demands have a uniform distribution $U\,(1, 30)$; however, $n/10$ was multiplied by 3.

II. Random clients, as shown in **Table 8**, belong to the Augerat set A in 1995 and specify that the coordinates are points between $[0,100] \times [0,100]$ placed randomly. The demands have a uniform distribution $U\,(1, 30)$; however, $n/10$ was multiplied by 3.

III. Clustered and random clients, as shown in **Table 9**, belong to the Augerat set X in 1995 and specify that the coordinates are points between $[0,1000]$ x $[0,1000]$ that are grouped and placed randomly, where k is the minimum feasible number of vehicles.

	Instances	Cap	n	k	Opt
Clustered	B-n31-k5	100	30	5	672
	B-n34-k5	100	33	5	788
	B-n35-k5	100	34	5	955
	B-n38-k6	100	37	6	805
	B-n39-k5	100	38	5	549
	B-n41-k6	100	40	6	829
	B-n43-k6	100	42	6	742
	B-n44-k7	100	43	7	909
	B-n45-k5	100	44	5	751
	B-n45-k6	100	44	6	678

Table 7.
Instances of set B.

	Instances	Cap	n	k	Opt
Random	A-n32-k5	100	31	5	784
	A-n33-k6	100	32	6	742
	A-n34-k5	100	33	5	778
	A-n36-k5	100	35	5	799
	A-n37-k5	100	36	5	669
	A-n37-k6	100	36	6	949
	A-n38-k5	100	37	5	730
	A-n39-k5	100	38	5	822
	A-n39-k6	100	38	6	831
	A-n44-k6	100	43	6	937

Table 8.
Instances of set A.

	Instances	Cap	n	k	Opt
Clustered and random	X-n101-k25	206	100	25	27,591
	X-n106-k14	600	105	14	26,362
	X-n110-k13	66	109	13	14,971
	X-n115-k10	169	114	10	12,747
	X-n120-k6	21	119	6	13,332
	X-n125-k30	188	124	30	55,539
	X-n129-k18	39	128	18	28,940
	X-n134-k13	643	133	13	10,916
	X-n139-k10	106	138	10	13,590
	X-n143-k7	1190	142	7	15,700

Table 9.
Instances of set X.

6. Results

The results for the 30 selected test cases are shown below, applying the heuristics and metaheuristics studied in Chapter 3 and 4 to know which one has a response that is closer or equal to the optimum by group of client positioning.

To define which has a better quality solution, the gap analysis or difference analysis is used, which consists in calculating the difference between the optimal solution and the solution obtained, divided for the solution obtained and expressed as a percentage.

The solution of the real case is also presented through the heuristics and metaheuristics that offer the best solution given the characteristic of the clients' positions.

6.1 Test cases resolved by heuristics

Clarke & Wright heuristics have better quality solutions, solving problems where customers with a small n are grouped.

Table 10 shows that for the group of clients with gathered positions, the gap is 3.63%; for the positions of random clients, the gap is 5.18%; and in less effective way for customers with grouped and random positions, it has a gap of 6.55%.

	Data						Clarke & Wright	
	Instances	Cap	*n*	*k*	Opt	*k*	Result	Gap (%)
Clustered	B-n31-k5	100	30	5	672	5	681.20	1.37
	B-n34-k5	100	33	5	788	5	794.30	0.80
	B-n35-k5	100	34	5	955	5	978.30	2.44
	B-n38-k6	100	37	6	805	6	832.10	3.37
	B-n39-k5	100	38	5	549	5	566.70	3.22
	B-n41-k6	100	40	6	829	7	898.10	8.34
	B-n43-k6	100	42	6	742	6	782.00	5.39
	B-n44-k7	100	43	7	909	7	937.70	3.16
	B-n45-k5	100	44	5	751	5	757.20	0.83
	B-n45-k6	100	44	6	678	7	727.80	7.35
	Average							3.63
Random	A-n32-k5	100	31	5	784	5	843.70	7.61
	A-n33-k6	100	32	6	742	7	776.30	4.62
	A-n34-k5	100	33	5	778	6	810.40	4.16
	A-n36-k5	100	35	5	799	5	828.50	3.69
	A-n37-k5	100	36	5	669	5	707.80	5.80
	A-n37-k6	100	36	6	949	6	976.60	2.91
	A-n38-k5	100	37	5	730	6	768.10	5.22
	A-n39-k5	100	38	5	822	5	902.00	9.73
	A-n39-k6	100	38	6	831	6	863.10	3.86
	A-n44-k6	100	43	6	937	7	976.00	4.16
	Average							5.18
Clustered and random	X-n101-k25	206	100	25	27,591	28	28940.00	4.89
	X-n106-k14	600	105	14	26,362	14	27280.00	3.48
	X-n110-k13	66	109	13	14,971	13	15870.00	6.00
	X-n115-k10	169	114	10	12,747	11	13490.00	5.83
	X-n120-k6	21	119	6	13,332	6	14540.00	9.06
	X-n125-k30	188	124	30	55,539	33	58830.00	5.93
	X-n129-k18	39	128	18	28,940	18	30300.00	4.70
	X-n134-k13	643	133	13	10,916	14	11520.00	5.53
	X-n139-k10	106	138	10	13,590	11	14530.00	6.92
	X-n143-k7	1190	142	7	15,700	7	17770.00	13.18
	Average							6.55

Table 10.
Clarke & Wright heuristics results.

It also compares the number of vehicles (k) that were obtained when solving each case against the optimal solution, and it was obtained that 12 cases had a vehicle more than the optimum B-n41-k6, B-n45-k6, A-n33-k6, A-n34-k5, A-n38-k5, A-n44-k6, X-n106-k14, X-n115-k10, X-n134-k13, and X-n139-k10, and two cases had three more than the optimal vehicles X-n101-k25 and X-n125-k30.

Sweep heuristics are more effective in solving problems where customers with a small n are grouped together. However, the difference in the average gap between random and grouped customers is short.

Table 11 shows that for the group of customers with grouped positions, the gap is 8.68%; for random customer positions; the gap is 8.85%; and in a less effective way for customers with grouped and random positions, it has a gap of 17.00%.

It also compares the number of vehicle numbers (k) that were obtained when solving each case against the optimal solution, and it was obtained that seven cases had a vehicle more than the optimal B-n45-k6, A-n38-k5, A-n39-k5, X-n115-k10, X-n129-k18, X-n134-k13, and X-n139-k10; one case had five more vehicles than the optimum X-n101-k25, and one case had six more vehicles than the optimal X-n125-k30.

In each group of clients, the sweep heuristic obtained better answers than the Clarke & Wright heuristics with 30% in the group of clients with a grouped position, 30% in the group of clients with a random position, and 20% in the group of clients with grouped and random position. In other words, Clarke & Wright heuristics are superior with 70% in the first two groups of clients and with 80% in the last group of clients.

A comparison among the values of the Distance traveled in the solution found by the heuristic, the optimal solution and the GAP for each one of the considered test instances is shown in **Table 12**.

On the other hand, **Table 13** shows a summary of the minimum, maximum and average gap for each of the three classes of problems considered: Clustered, Random and Clustered, and Random.

6.2 Test cases resolved through metaheuristics

The GRASP metaheuristics are based on a previous solution for which Clarke & Wright heuristic responses were selected since their responses are of better quality than the sweep heuristics.

The following parameters were considered for its implementation:

- $\alpha = 0.5$

- Number of iterations = 10,000

GRASP's metaheuristics are more effective in solving problems where customers with a small n are grouped together.

Customers	Minimum gap (%)	Maximum gap (%)	Average gap (%)
Clustered	0.80	8.34	3.63
Random	2.91	9.73	5.18
Clustered and random	3.48	13.18	6.55

Table 11.
Clarke & Wright heuristic gaps comparison.

		Data					Sweep		
	Instances	Cap	*n*	*k*	Opt	*k*	Result	Gap (%)
Clustered	B-n31-k5	100	30	5	672	5	696.69	3.67
	B-n34-k5	100	33	5	788	5	889.51	12.88
	B-n35-k5	100	34	5	955	5	966.93	1.25
	B-n38-k6	100	37	6	805	6	838.99	4.22
	B-n39-k5	100	38	5	549	5	613.45	11.74
	B-n41-k6	100	40	6	829	6	884.53	6.70
	B-n43-k6	100	42	6	742	6	750.92	1.20
	B-n44-k7	100	43	7	909	7	1137.46	25.13
	B-n45-k5	100	44	5	751	5	836.08	11.33
	B-n45-k6	100	44	6	678	7	736.62	8.65
	Average							8.68
Random	A-n32-k5	100	31	5	784	5	885.04	12.89
	A-n33-k6	100	32	6	742	6	751.65	1.30
	A-n34-k5	100	33	5	778	5	786.44	1.08
	A-n36-k5	100	35	5	799	5	862.71	7.97
	A-n37-k5	100	36	5	669	5	736.35	10.07
	A-n37-k6	100	36	6	949	7	1087.46	14.59
	A-n38-k5	100	37	5	730	6	818.46	12.12
	A-n39-k5	100	38	5	822	5	882.53	7.36
	A-n39-k6	100	38	6	831	6	900.14	8.32
	A-n44-k6	100	43	6	937	6	1056.84	12.79
	Average							8.85
Clustered and random	X-n101-k25	206	100	25	27,591	30	34368.50	24.56
	X-n106-k14	600	105	14	26,362	14	30035.90	13.94
	X-n110-k13	66	109	13	14,971	13	15769.30	5.33
	X-n115-k10	169	114	10	12,747	11	14894.20	16.84
	X-n120-k6	21	119	6	13,332	6	14495.40	8.73
	X-n125-k30	188	124	30	55,539	36	69342.40	24.85
	X-n129-k18	39	128	18	28,940	19	36941.80	27.65
	X-n134-k13	643	133	13	10,916	14	13835.90	26.75
	X-n139-k10	106	138	10	13,590	11	14850.90	9.28
	X-n143-k7	1190	142	7	15,700	7	17593.50	12.06
	Average							17.00

Table 12.
Sweep heuristic results.

Table 14 shows that for the clients with grouped positions; the gap is 3.09%; for the positions of random clients, the gap is 4.38%; and less effectively for customers with grouped and random positions, it has a gap of 5.97%.

Customers	Minimum gap (%)	Maximum gap (%)	Average gap (%)
Clustered	1.20	25.13	8.68
Random	1.08	14.59	8.85
Clustered and random	5.33	27.65	17.00

Table 13.
Sweep heuristic gap comparison.

		Data					Grasp	
	Instances	cap	*n*	*k*	Opt	*k*	Result	Gap (%)
Clustered	B-n31-k5	100	30	5	672	5	679.05	1.05
	B-n34-k5	100	33	5	788	5	788.00	0.00
	B-n35-k5	100	34	5	955	5	968.85	1.45
	B-n38-k6	100	37	6	805	6	830.45	3.16
	B-n39-k5	100	38	5	549	5	564.85	2.89
	B-n41-k6	100	40	6	829	7	897.24	8.23
	B-n43-k6	100	42	6	742	6	777.98	4.85
	B-n44-k7	100	43	7	909	7	932.36	2.57
	B-n45-k5	100	44	5	751	5	755.23	0.56
	B-n45-k6	100	44	6	678	7	719.80	6.17
	Average							3.09
Random	A-n32-k5	100	31	5	784	5	830.67	5.95
	A-n33-k6	100	32	6	742	7	776.02	4.58
	A-n34-k5	100	33	5	778	6	809.38	4.03
	A-n36-k5	100	35	5	799	5	823.20	3.03
	A-n37-k5	100	36	5	669	5	695.42	3.95
	A-n37-k6	100	36	6	949	6	976.61	2.91
	A-n38-k5	100	37	5	730	6	765.87	4.91
	A-n39-k5	100	38	5	822	5	901.99	9.73
	A-n39-k6	100	38	6	831	6	856.93	3.12
	A-n44-k6	100	43	6	937	7	951.73	1.57
	Average							4.38
Clustered and random	X-n101-k25	206	100	25	27,591	28	28891.90	4.71
	X-n106-k14	600	105	14	26,362	14	27199.80	3.18
	X-n110-k13	66	109	13	14,971	13	15847.90	5.86
	X-n115-k10	169	114	10	12,747	11	13436.60	5.41
	X-n120-k6	21	119	6	13,332	6	14192.90	6.46
	X-n125-k30	188	124	30	55,539	33	58809.10	5.89
	X-n129-k18	39	128	18	28,940	18	30298.40	4.69
	X-n134-k13	643	133	13	10,916	14	11492.20	5.28
	X-n139-k10	106	138	10	13,590	11	14521.10	6.85
	X-n143-k7	1190	142	7	15,700	7	17491.80	11.41
	Average							5.97

Table 14.
GRASP metaheuristic results.

On average metaheuristic GRASP based is better than Clarke & Wright heuristics by 0.53%, 0.77%, and 0.38% in the solutions of the positions of the grouped, random, and grouped and random clients, the group of clients with random positions being those that obtained a greater improvement in the quality of the solutions.

		Data					**Simulated annealing**	
	Instances	**Cap**	**n**	**k**	**Opt**	**k**	**Result**	**Gap (%)**
Clustered	B-n31-k5	100	30	5	672	5	681.20	1.37
	B-n34-k5	100	33	5	788	5	793.20	0.66
	B-n35-k5	100	34	5	955	5	959.50	0.47
	B-n38-k6	100	37	6	805	6	819.50	1.80
	B-n39-k5	100	38	5	549	5	565.00	2.91
	B-n41-k6	100	40	6	829	7	897.00	8.20
	B-n43-k6	100	42	6	742	6	778.60	4.93
	B-n44-k7	100	43	7	909	7	937.30	3.11
	B-n45-k5	100	44	5	751	5	756.20	0.69
	B-n45-k6	100	44	6	678	7	726.16	7.10
	Average							3.13
Random	A-n32-k5	100	31	5	784	5	830.70	5.96
	A-n33-k6	100	32	6	742	7	776.30	4.62
	A-n34-k5	100	33	5	778	6	810.40	4.16
	A-n36-k5	100	35	5	799	5	828.50	3.69
	A-n37-k5	100	36	5	669	5	695.00	3.89
	A-n37-k6	100	36	6	949	6	976.60	2.91
	A-n38-k5	100	37	5	730	6	762.00	4.38
	A-n39-k5	100	38	5	822	5	888.60	8.10
	A-n39-k6	100	38	6	831	6	856.90	3.12
	A-n44-k6	100	43	6	937	7	967.60	3.27
	Average							4.41
Clustered and random	X-n101-k25	206	100	25	27,591	28	28850.00	4.56
	X-n106-k14	600	105	14	26,362	14	27240.00	3.33
	X-n110-k13	66	109	13	14,971	13	15790.00	5.47
	X-n115-k10	169	114	10	12,747	11	13480.00	5.75
	X-n120-k6	21	119	6	13,332	6	14420.00	8.16
	X-n125-k30	188	124	30	55,539	33	58790.00	5.85
	X-n129-k18	39	128	18	28,940	18	30300.00	4.70
	X-n134-k13	643	133	13	10,916	14	11500.00	5.35
	X-n139-k10	106	138	10	13,590	11	14530.00	6.92
	X-n143-k7	1190	142	7	15,700	7	17670.00	12.55
	Average							6.26

Table 15.
Results of simulated annealing metaheuristics.

The simulated annealing metaheuristics start from a previous solution for which Clarke & Wright heuristic responses were selected since their responses are of better quality than the sweep heuristics.

The following parameters were considered for its implementation:

- Current temperature = 250

- Final temperature = 10

- Cooling coefficient = 0.8

- Number of iterations = 10,000

The simulated annealing metaheuristic is more effective in solving problems where customers with a small n are grouped together.

Table 15 and **17** shows that for clients with grouped positions, the gap is 3.13%; for the positions of random clients, the gap is 4.41%; and less effectively for clients with grouped and random positions, it has a gap of 6.26%.

On average simulated annealing heuristics based is better than Clarke & Wright heuristics by 0.52%, 0.71%, and 0.21% in the solutions of grouped, random, and grouped and random clients' positions, the group of clients with random positions being those that obtained a greater improvement in the solutions quality.

Within each group of clients, the simulated annealing metaheuristics obtained better answers than the GRASP metaheuristics with 30% in the group of clients with a grouped position, 50% in the group of clients with a random position, and 40% in the group of clients with grouped and random position. That is, the GRASP metaheuristic is superior with 70% in the first group and with 60% in the third group of clients and is equal with 50% in the second group of clients.

A comparison between the minimum and maximum gap within each group is established in **Table 16** test results with clustered clients have better results. Therefore, obtaining a minimum gap of 0%, that is, in the case of B-n34-k5, the GRASP metaheuristic obtained the optimal solution (**Table 17**).

Customers	Minimum gap (%)	Maximum gap (%)	Average gap (%)
Clustered	0.00	8.23	3.09
Random	1.57	9.73	4.38
Clustered and random	3.18	11.41	5.97

Table 16.
GRASP metaheuristic gap comparison.

Customers	Minimum gap (%)	Maximum gap (%)	Average gap (%)
Clustered	0.47	8.20	3.13
Random	2.91	8.10	4.41
Clustered and random	3.33	12.55	6.26

Table 17.
Simulated annealing metaheuristics gap comparison.

7. Conclusions

The results obtained by solving the test cases by heuristics and metaheuristics show both generate better quality solutions when solving cases where customers are grouped together and had their worst result in the group of clients with a grouped and random position since they had large n with reference to the other groups of clients. Between the heuristics, Clarke & Wright heuristics generated better quality results than sweep heuristics, having a big difference in the maximum gaps of each one for each group of clients. However, the numbers of vehicles obtained in the solutions of both heuristics were compared with the optimal solution, and the sweep heuristics had more solutions in which it reached the optimum. This is a very important contribution, since Clarke & Wright heuristics get solutions with shorter distances than the sweep heuristic, but this gets greater distances with fewer vehicle units. For some companies it will be more important to reduce the units to buy than the distance traveled.

Analyzing the metaheuristics, the GRASP metaheuristics generated better quality results than simulated annealing metaheuristics, with minimal differences in average gap for each group of clients, and both metaheuristics obtained greater improvements in relation to the initial solutions of Clarke & Wright heuristics in the test cases of randomized clients. Also GRASP algorithm with B-n34-k5 case of grouped customers reached the optimal solution, being the only test instance of the thirty that were done.

It is recommended that for future studies, each group of clients by positioning has a number of clients (n) with greater variability to be able to deduce exactly if heuristics and metaheuristics have better or worse solutions when n are larger or smaller.

Author details

Fernando Francisco Sandoya Sánchez, Carmen Andrea Letamendi Lazo[*]
and Fanny Yamel Sanabria Quiñónez
Escuela Superior Politécnica del Litoral (ESPOL), Guayaquil, Ecuador

*Address all correspondence to: cletamendi@gmail.com

IntechOpen

References

[1] Ballou, RH. Business Logistics/ Supply Chain Management: Planning, Organizing, and Controlling the Supply Chain. Upper Saddle River, New Jersey: Pearson/Prentice Hall; 2004 (Print)

[2] Toth P, Vigo D, editors. The Vehicle Routing Problem. Society for Industrial and Applied Mathematics; Jan 2002. Available from: http://dx.doi.org/ 10.1137/1.9780898718515

[3] Braekers K, Ramaekers K, Van Nieuwenhuysec I. The vehicle routing problem: State of the art classification and review. Computers & Industrial Engineering. 2016;**99**:300-313

[4] Safra S, Schwartz O. On the complexity of approximating TSP with neighborhoods and related problems. Computational Complexity. 2005;**14**: 281-307

[5] Abdel-Basset M, Abdel-Fatah L, Kumar Sangaiah A. Metaheuristic algorithms: A comprehensive review. Computational Intelligence for Multimedia Big Data on the Cloud with Engineering Applications. 2018:185-231

[6] Sandoya F, Martinez-Gavara A, Aceves R, Duarte A, Martı R. Diversity and equity models. In: Martí R, Panos P, Resende M, editors. Handbook of Heuristics. Cham: Springer; 2015. pp. 1-20

[7] Altınel I, Öncan T. A new enhancement of the Clarke and Wright savings heuristic for the capacitated vehicle routing problem. Journal of the Operational Research Society. 2005; **56**(8):954-961

[8] Resende M, Ribeiro C. Optimization by GRASP. Greedy Randomized Adaptive Search Procedures. New York, NY: Springer; 2016

[9] Marti R, Sandoya F. GRASP and path relinking for the equitable dispersion problem. Computers & Operations Research. 2013;**40**(12):3091-3099

[10] Augerat. Capacitated VRP Instances. 2013. Available from: http://neo.lcc.uma.es/vrp/vrp-instances/capacitated-vrp-instances/ [Accessed: 12 September 2019]